The Elusive Bonanza

The Elusive Bonanza

The Story of Oil Shale—
America's Richest and Most Neglected
Natural Resource

by Chris Welles

 E. P. Dutton & Co., Inc.   New York   1970

Published simultaneously in Canada by Clarke, Irwin & Company Limited, Toronto and Vancouver

Library of Congress Catalog Card Number: 74-122780

SBN: 0-525-09761-9

To My Father

Contents

Acknowledgments

Perhaps the most challenging, as well as the most exasperating aspect of writing this book has been that the subject of oil shale is almost as unexplored as the oil shale lands themselves. In fact, to my knowledge, this is the first reasonably comprehensive book to be written on oil shale in 50 years. There is a multitudinous amount of questions, problems, and enigmas but only a meager amount of answers, solutions, and certainties. The almost mind-bending complexity of oil shale surpasses anything I have ever encountered, and whatever its other attributes, good and bad, this book, I think, will not be easily skimmed. Its purpose is not to promulgate final judgments but to offer tentative interpretations which, hopefully, will stimulate further investigation into one of this nation's richest and most neglected natural resources.

My principal source of information has been nearly two hundred interviews with oil industry executives, Interior Department officials, congressmen, economists, lawyers, geologists, engineers, prospectors, and many other people whose only common denominator is that they have been watching the shale lands for a long time to see if anything was ever going to happen to them. A number of published sources were of considerable assistance to me. The most voluminous information on oil shale ever assembled in one place is the records of Contests #359 and #360 by the Bureau of Land Management of the Department of the Interior concerning the validity of some mining claims on federal oil shale land. Since a chief issue in these contests is the economic feasibility of producing oil from oil shale, the testimony and documents proved invaluable in shedding light on the question which is the central concern of this book: Why has the Colorado oil shale never been commercially developed? Extremely useful also have been hearings on shale by the Senate Committee on Interior and Insular Affairs in May, 1965, and February and September, 1967, and by the Senate Subcommittee on Antitrust and Monopoly in April and May, 1967. I have drawn particularly on the testimony and other writings of Dr. Henry Steele of the University of Houston and Dr. Walter J. Mead of the University of California at Santa Barbara. The most convoluted issues regarding oil shale concern the legal status of the shale lands, and I received excellent guidance from *Legal Study of Oil Shale on Public Lands,* prepared by the University of Denver College of Law under the guidance of Prof. Gary L. Widman for the Public Land Law Review Commission in April, 1969. Much helpful information on current technology, legal developments, and other areas is contained in the records of the Annual Symposiums on Oil Shale sponsored by the Colorado School of Mines and the American Institute of Mining, Metallurgical, and Petroleum Engineers and published in the *Quarterly of the Colorado School of Mines.*

I would like to extend special gratitude to Mrs. Nancy Belliveau, who did a great deal of research work for me, and to J. R. Freeman, a courageous and crusading newspaper editor. I have come to disagree with many of his theories, but I have nothing but respect for his zealous devotion to uncovering the truth about oil shale. Without his initial assistance I never would have written this book.

One additional note: following the main body of this book is an Appendix concerning some of the circumstances of my personal involvement with the subject of oil shale while I was Business Editor of *Life* magazine. Usually I would consider such matters of little concern to the reader. But in this case there has been public controversy about aspects of my investigations which could influence the way some people evaluate the point of view of this book. I therefore feel it necessary to relate the circumstances, my interpretation of them, and the statements of some of the other people who were involved.

Chris Welles

Dumont, New Jersey
February, 1970

Introduction

The oil industry in the United States is a "government-created, government-supported, and government-subsidized cartel," Dr. Walter Adams, president of Michigan State University and an economist, told the Senate Antitrust and Monopoly Subcommittee during hearings on governmental restrictions of petroleum imports in April, 1969. He said the industry could be explained best through theories held by the late Professor Joseph A. Schumpeter, who asserted that

. . . the capitalist process was rooted not in classical price competition, but rather the competition from the new commodity, the new technology, the new source of supply, the new type of organization—competition which commands a decisive cost or quality advantage and which strikes not at the margin of the profits and outputs of existing firms, but at their very foundations and their very lives. The very essence of capitalism, according to Schumpeter, was the perennial gale of creative destruction in which existing power positions and entrenched advantage were constantly displaced by new organizations and new power complexes. This gale of creative destruction was to be not only the harbinger of progress, but also the built-in safeguard against the vices of monopoly and privilege.

What was obvious to Schumpeter and other analysts of economic power, was also apparent to those who might suffer from the gales of change. They quickly and instinctively understood that storm shelters had to be built to protect themselves against this destructive force. The mechanism which was of undoubted public benefit carried with it exorbitant private costs. And, since private storm shelters in the form of cartels and monopolies were either unlawful, unfeasible, or inadequate, they have turned increasingly to Government for succor and support. By manipulation of the state for private ends, the possessors of entrenched power have found the most felicitous instrument for insulating themselves against, and immunizing themselves from, the Schumpeterian gale.

Economic history shows that it is quite characteristic for large, well-established corporations and industries to resist change, to become less and less dynamic, entrepreneurial and innovative and more and more repressive, protective and ossified. The reason is simple: they become committed to, and define themselves in terms of existing ways of doing things. In *Technology and Change* (Delacorte Press, 1967), Dr. Donald A. Schon analyzed three "mature" industries—building, textiles and machine tools—and concluded that the major innovations have come not from companies traditionally associated with those industries but from independent inventors, foreign technology, start-up of new small firms and invasion of the industry by firms from other industries. It was not GE or RCA that pioneered with the computer but IBM, not the old printing companies that developed the new reproduction and duplication techniques but Xerox.

As this book demonstrates, the oil industry, with the help of state and federal governments, is distinguished by an unusual desire and ability to withstand the Schumpeterian gale and alteration of the comfortable status quo. Despite its oft-expressed reverence for free enterprise, no other industry, in fact, can match its use of public institutions for the protection of its private self-interest. Historically, it has faced two forms of potentially disruptive change. One is competition from other energy sources. Because of the huge capital expenditures involved and other factors, broad shifts in energy consumption patterns occur gradually. But the industry, virtually unchallenged by antitrust authorities, has been actively hedging itself by obtaining broad interests in natural gas, coal and nuclear power. The other is excess supplies of crude oil. With the assistance of a dense complex of governmental regulations, the industry has been quite effective in restraining supply to match demand—thus permitting maintenance of unnaturally high crude prices—and in ensuring that crude flows through conventional industry channels.

The true Achilles' heel of the oil cartel is a sizable source of supply unaffected by existing controls. This supply would be especially dangerous if it consisted of a substance that was more than a mere duplicate of crude oil, a substance that was technologically superior, more attractive to the marketplace and less expensive to produce. This substance might not only cause crude prices to drop but might even start replacing crude oil altogether.

Oil shale represents just such an Achilles' heel to the petroleum industry. It is a black, organic rock which when heated releases a form of oil which can be easily refined to produce any conventional petroleum product, from gasoline to fuel oil to asphalt to petrochemicals. The oil shale in one deposit in particular—located in western Colorado—is so incredibly abundant that if even a small percentage were turned into shale oil, the supply would be sufficient to meet the nation's petroleum needs well beyond the year 2000—indeed probably until all fossil fuels become obsolete. Though this deposit has never been commercially exploited to any meaningful extent, the evidence is overwhelming that production of shale oil would be not only profitable but a good deal more profitable than exploration for and production of new domestic reserves of crude oil.

To oil men, shale oil represents a vast new potential source of supply which, depending on the circumstances, they might have a quite limited ability to control or restrain. More generally, it represents a change, something new and different, an unknown. To a company in a competitive market, such a substance would mean a possible opportunity to improve its position. To a member of a cartel, whose executives typically sink into a kind of protective sloth, a sterile lassitude, it means a possible threat or, at the very least, something not worth making the effort to bother with. Consequently, the oil industry has generally resisted or refused to engage in any significant efforts to develop the shale or even to determine how economical shale oil production might be.

The federal government, which as owner of over 80 percent of the Colorado shale has broad powers to determine its use, has docilely acquiesced to the oil industry's point of view. Not only have government officials been unwilling to assume the political risks of stimulating development of such a valuable resource, but they have naively assumed that what's good for the oil industry is good for the country, that if the shale were worth developing, the oil industry would be trying to develop it. This is a little like judging the economic value of the aluminum soft drink can by whether or not the steel industry was making it. The government has yet to produce a meaningful oil shale development policy, and has not conducted or sponsored anything more than minuscule research on oil shale in over 15 years. The shale lands today remain virtually untouched.

All of this would be rather academic if shale oil production did not offer such great potential benefits for consumers and the nation in general. At a time when, despite recent Alaskan discoveries, we appear to be running very short of domestic crude oil reserves, shale oil could assure us of a lasting domestic fuel supply, and we would no longer face increasing dependence on supplies from the unstable Middle East and other overseas areas, with the threat to our balance of payments and our national security that such dependence implies. If shale oil production were to prove as economical as many studies predict, the price of all petroleum products could eventually be substantially reduced, with annual savings to consumers in the billions of dollars. Embodying the latest safeguards, shale oil production facilities could be almost pollution free compared to the serious environmental hazards of today's petroleum industry, such as rupture of offshore wells and spillage from oceangoing tankers.

Whether these benefits come to pass depends not only on whether the shale is developed but, if it is developed, who the developers are. Frequently when an entrenched, cartel-like industry has been threatened from the outside by a technological advance or a superior new product whose eventual introduction appears unavoidable, the industry has modified its policy of massive resistance to include steps to obtain control of the advance or product and thus mollify its possibly disruptive effects. An excellent example is cable television which, because of its ability to introduce cheaply into a local viewing area a large number of distortion-free channels, poses a serious long-term threat to the structure and profitability of the broadcasting industry. Initially, broadcasters were unremittingly hostile to CATV. Only after hundreds of CATV systems were constructed by independent entrepreneurs and it became obvious that CATV would continue to proliferate, did broadcasters—still not ceasing their efforts to restrain the interloper—begin getting into the CATV business themselves. Because the ultimate regulatory power over the airwaves rests with the Federal Communications Commission, the outcome of the struggle is in doubt. But as larger and larger slices of the CATV industry are acquired by broadcasting companies, the long-run probability of disruption—and socially useful change—declines.

There are currently signs that some oil companies have begun to take a view toward oil shale somewhat similar to that of the broadcasters toward CATV. Due to the likelihood of increasing pressure for shale development caused by dwindling domestic crude reserves, oil companies have already bought up most of the privately owned shale land, and a few of the more adventurous firms have recently acquired a sizable stake in existing oil shale technology. If this trend continues and shale oil production by oil companies commences, the country will enjoy the benefits only to the degree, if at all, that the oil industry deems such benefits to be in its own self-interest. The fact that such a large portion of this resource is still owned by the people gives the nation an opportunity to ensure that its development will instead be in the public interest. The opportunity exists, in fact, to accomplish what the Justice Department has been unable to despite 50 years of trying: to introduce a free, open, competitive market into the country's most monopolistic industry. Unfortunately, there is a very real current danger that instead the government will continue to protect the oil industry from the Schumpeterian gale, and will lease or otherwise hand over control of the public shale lands to the major oil companies with only minimal provisions to protect the public interest. It is not a happy prospect. As Michigan Senator Philip A. Hart has warned, "Monopolization of this enormous asset would be a national tragedy."

PART ONE

"We are a people with a vast treasure,
unaware of its existence."
> —John Kenneth Galbraith,
> Professor of Economics,
> Harvard University,
> April 18, 1967

Chapter 1:
There is a lot of oil
in the shale out in Colorado,
but for some reason nobody is doing
anything much about it.

In the small, worn, dusty towns on the banks of the meandering Colorado River 200 miles west of Denver, there are men with hard faces and long gazes. They resemble the storied Western prospectors who imagine a glistening vein will be uncovered with the very next swing of the pick. But their eyes do not have the same hopefulness that helps to drive out the frustrations and hardships from previous swings of the pick. In the eyes and in the voices of the men on the banks of the Colorado, frustration and hardship, as well as exhaustion and even anger are all too manifest and explicit. For unlike other seekers of mineral wealth, they know precisely where the treasure is and how to get it. It is tantalizingly visible, a 75-foot-thick, whitened horizontal streak stretching under the rim of the nearby Roan Cliffs which loom 4,000 feet above the river level. The whitened streak is the thin edge of a vast, much thicker underground deposit of a substance known as "oil shale," an organic black rock which when heated can be converted into petroleum and gasoline. Though few Americans have ever heard of oil shale, this deposit, which extends from western Colorado into eastern Utah and southern Wyoming, has been called by former Secretary of the Interior Stewart L. Udall "the largest untapped source of hydrocarbon energy known to the world." Albert C. Rubel, a geologist and former president of the Union Oil Company of California, termed it "the greatest package of potential energy on the face of the globe." The statistics of its extent are almost beyond comprehension. If all of the shale in this one deposit were converted to oil, the result would be, according to a

1965 U.S. Geological Survey study, *eight trillion barrels* (one barrel = 42 U.S. gallons), 20 times the world's proven reserves of petroleum and enough to supply the United States, at its present rate of consumption, for nearly 2,000 years. Excluding thin, lean portions of the deposit, the oil from this shale is worth potentially between $5 and $10 trillion, or more than $25,000 for every man, woman, and child in America.

The men living along the Colorado River, who call themselves "oil shalers," have been waiting many, many years for the high cliffs to yield up some of these riches, to shower their small communities with dollars, to drive up real estate values, to create giant industries, to bless them forever with unimaginable bliss. For a century, these men and their fathers and their grandfathers have eagerly consumed a plethora of reports, studies, and analyses which have waxed to levels of ecstasy about the resource's incredible value and have all but promised that legal tender would emanate in abundance in just a few years. "Within the decade," the phrase usually went. A typical recent study by the United Nations Department of Economic and Social Affairs in 1965 said production of oil from shale was fast becoming a "technical and economic reality. . . . Once this point is reached, a new industry will emerge that promises to dwarf petroleum in meeting the world's energy needs."

But nothing is happening. Except for a few cattle ranchers and deer hunters the land is all but deserted. No one is selling shale oil. There is no stampede to build factories, to cash in, to get rich. The dusty towns, in fact, are dying. Industry is moving away. Stores are closing. The young do not return after college. The $10 trillion worth of oil shale remains virtually untouched.

The answer to why this is so will be the chief concern of this book.

It is first necessary to understand just what oil shale is. Actually, it is neither oil nor shale. According to one geological definition, it is a "fine-textured sedimentary rock containing organic matter that yields substantial amounts of oil and hydrocarbon gas by destructive distillation." The source of the energy in oil shale is very similar to that in such other "fossil fuels" as petroleum and coal: the remains of plants and animals which after dying were buried beneath the ground. A dead plant or animal exposed to the air will rot or oxidize, a burning process that consumes the organism's carbon and hydrogen, thus using up its potential energy. But trapped underground, organic material cannot fully oxidize, and the carbon and hydrogen are preserved, so they may be oxidized and converted into energy by man.

The form these underground remains take is dependent on geological and climatic conditions. Petroleum is generally created at the bottom of present and former oceans by plant and animal matter trapped in layers of sedimentary rock. Succeeding layers over millions of years exert sufficient heat and pressure to distill from the matter petroleum and natural gas—the greater the heat and pressure, the lighter the distillate and the closer it is to dry gas. Petroleum often forms underground pools in anomalies and gaps in the earth's stratification or permeates layers of porous rock. In both cases it can be tapped by surface wells.

Coal was formed principally from plant life in land areas. About 200–250 million years ago, during a period of extremely lush vegetation and extensive swamps, generations of decaying growths formed a kind of mulch which was compressed and hardened by successive layers of earth and plant material. When the swamps evaporated, the mulch dried and formed black rock.

In the opinion of many geologists, oil shale, though bearing some resemblances to coal, is really incompletely developed petroleum. Formation of oil shale deposits began, like petroleum, with the accumulation of dying plant and animal matter in bodies of water, usually inland seas and lakes. The organic material mixed with sediments and settled to the water's bottom, but due to insufficient heat and pressure, actual distillation did not occur. Created instead was "kerogen," a rubbery solid about 80 percent carbon and 10 percent hydrogen which mixed chemically with the sediments to form a tough, dense, inelastic, insoluble rock, whose color ranges from light tan to black depending on the percentage of kerogen. The delineation of the layers is often clearly visible.

The chief chemical difference between kerogen and petroleum is geometric: unlike petroleum the kerogen molecules are connected with elaborate chains. To convert kerogen to a close equivalent of petroleum, one need only apply the heat it did not receive from the earth. At a temperature between 850 and 900 degrees F. "retorting" occurs. The molecular chains break, and the kerogen is distilled into, by weight, about 66 percent shale oil (which is usually released as a vapor), 25 percent coke-like solid matter, and 9 percent combustible gas. Thin layers of shale with a high kerogen content can be retorted with a match, which creates a sooty flame and an acrid odor. The amount of shale oil produced varies, of course, with the percentage of kerogen, ranging from less than five gallons per ton of shale rock to nearly 100 gallons per ton. One cubic foot of reasonably rich shale will yield about three gallons of shale oil. The biggest residue from retorting is "spent shale" which weighs only slightly less than the original shale rock and expands during retorting to a volume between 20 and 40 percent greater, meaning that more space is required to dispose of these tailings than the hole from which the shale rock was dug.

Shale oil has important physical differences from petroleum. Depending on the retorting method, it usually has a high viscosity due to a large wax content, and at room temperature it is so thick and tar-like it will not flow. To be moved through a pipeline it must be either heated to over 90 degrees F. or further "cracked" or distilled. Shale oil also contains excessive impurities including sulphur, tars, and such gases as nitrogen and ammonia. But once they are removed, shale oil can be refined easily into a full range of petroleum products from fuel oil and asphalt to gasoline and petrochemicals.

Deposits of shale occur throughout the world on every continent, constituting, in fact, a greater potential energy resource than any other natural material except coal—far greater than petroleum, natural gas, uranium, and hydro power. Significant amounts are known to exist in Brazil, Scotland, Sweden, China, Russia, Australia, Germany, France, Canada, and Thailand. At least 30 states contain shale deposits, and in wide areas just west of the Appalachian Mountains, much oil shale is mixed with coal.

By far the richest and most extensive oil shale deposit in the world is located in what is called the Green River Formation. This is contained in a roughly circular area of about 17,000 square miles extending out about 150 miles from where Colorado, Utah, and Wyoming intersect and through which the Green River, one of the principal tributaries of the Colorado River, flows. About 72 percent of this land, containing between 80 and 85 percent of the shale, is owned by the federal government.

The staggering dimensions of this resource were created by a huge inland sea which covered much of the western United States well over 60 million years ago. As the sea slowly evaporated, two large, rather shallow, fresh water lakes covering about 34,000 square miles remained. What geologists call Lake Gosiute covered southern Wyoming and was separated from Lake Uinta, which covered Utah and Colorado, by a wide plateau. The lakes were surrounded by high hills and frequent rains washed huge amounts of plant and animal life into the water. Meanwhile, the deep earth forces which were thrusting up the Rocky

Mountains raised the lakes, eventually as high as 9,400 feet above sea level. This "crustal upwarp" cut off outlets and sources and the lakes became stagnant. Broad swamps and mud flats developed along the shore lines, which were later to become coal deposits. The accumulating organic matter in the water acted as a kind of naural pollutant, not unlike the manmade variety of Lake Erie today in that it stimulated the growth of immense quantities of algae. Bacteria, fungi, protozoa and other microorganisms also flourished. Vast accumulations of plant spores, pollen, and other grains added to the debris. Over perhaps 10 million years of remarkably constant climatic conditions, this material sank to the bottom to form thick layers of ooze, along with sedimentary clay and sand with which it chemically combined. As the lakes continued to shrink, the ooze was covered with layers of volcanic ash and other rock torn loose from the convulsive forming of the nearby mountains. Though in some sections pools of conventional oil and gas were to form, the upwarp did not allow most of the organic material to progress beyond the kerogen stage.

The thickness of the resulting formation is between 1,800 and 3,500 feet. The richness and thickness of the layers of oil shale—the host rock is called marlstone, a close relative of limestone, which contains also various amounts of calcium, sodium, potassium, and several metalliferous substances—varies widely depending on the shape of the shore line. The particular climate during the time a particular layer was deposited and other factors affected plant and animal growth. Perhaps 80 percent of the Green River Formation's oil shale is located in the Piceance (pronounced pe'-änce) Creek Basin, the Colorado portion of Lake Uinta, which ranges over 1,380 square miles in Garfield and Rio Blanco Counties. Along its edges, the shale is only a few feet thick and close to the surface of the ground. But toward the basin's center, which lies about 25 miles southwest of Meeker, the shale layers incline downward and thicken, becoming about 2,300 feet thick and covered with a thousand feet of "overburden." There are two principal layers of rich shale. The upper is the "Mahogany Zone," so-termed because when polished, rock from this zone resembles mahogany. Formed during an especially productive 200,000-year epoch (each year of the Piceance Basin's accumulation produced a layer about $\frac{1}{100}$ of an inch thick) it lies fairly close to the surface and is from 50 to 200 feet thick. Mahogany shale contains usually at least 30 gallons of oil per ton. A lower shale zone is less rich than the Mahogany Zone but is much thicker and occurs principally at the center of the basin, where the Mahogany Zone itself is richest and broadest. A single square mile in the basin's center may contain 2½ billion barrels of shale oil, enough to supply the entire country for six months.

Exactly what sections of the shale of what richness and what thickness at what depth it is economical to recover is impossible to say. Obviously only the richest and most accessible shale would be of initial interest, and it is conceivable that even if a huge shale oil industry should spring up that the great bulk of the deposit would never be exploited. According to the Geological Survey, the Piceance Creek Basin contains between 450 and 500 billion barrels of oil in deposits at least 10 feet thick containing at least 25 gallons per ton—this would be about a hundred year's domestic supply at present consumption rates. Another 800 billion barrels are contained in deposits with a richness of between 10 and 25 gallons per ton. Wyoming shales are generally lean and thin, but the Utah shales, though relatively unexplored, are believed by some geologists to be nearly as rich as those in Colorado.

The surface of the land in the Piceance Creek Basin gives little hint of the potential wealth beneath. It is a barren and bleak plateau, broken by low, rolling hills and jagged canyons carved by small, intermittent creeks and streams which flow into the Colorado River. It is covered with sparse patches of sage, piñon, juniper, and a few occasional groves of pine, spruce, and aspen at the higher elevations. Sections of the basin are leased to ranchers for summer grazing—stock raising is the principal local industry—and a few rutted dirt roads wind over some areas. The only other inhabitants are some herds of migrant deer, elk, and antelope. The basin's shale might never have been discovered had it not been for the Colorado River, which sliced a deep, wide gorge through the southern tip of the basin, revealing, close to the top of the Roan Cliffs, the Mahogany "Ledge," as the "outcropping" of the Mahogany Zone is called. Exposed also is the Mahogany "Marker," a six-inch layer of volcanic ash which clearly identifies the top of the Zone. The Mahogany Ledge outcrops not only along the Colorado River but along the numerous canyons created by its tributaries, and virtually all of the shale oil research operations have used shale taken from cliff-face mines dug into the Mahogany Ledge.

The fact of the Piceance Creek Basin shale deposit, if not its overwhelming extent, has been known for centuries. The Ute Indians, who once inhabited the area, were often startled when bolts of lightning during summer storms caused flames to shoot from shale outcrops, and they called the shale "the rock that burns." Early pioneers were even more ignorant, and the probably apocryphal tale is told—and retold in virtually every article ever written about oil shale—about an individual named Mike Callahan who settled near Rifle and built himself an impressive house whose *pièce de résistance* was a giant fireplace built from the black rock he had found in the nearby cliffs. At a glorious housewarming party, Callahan proudly lit the fireplace's inaugural fire which quickly escalated into a roaring conflagration, leveling both house and fireplace, and perhaps even Callahan.

Production of oil from shale occurred in other areas long before interest developed in the Green River Formation. The earliest oil shale operation may have been in England, where a patent was granted in 1694 for a process to distill "oyle from a kind of stone." In the United States, shale oil production predated the inception of the petroleum industry by several decades. During the early 1800s, and possibly as early as Revolutionary times, lamp oil, lubricants, and medicinal substances were produced from the rather lean shales around the Appalachians. Around 1850 there was talk of a shale oil boom, a new answer to the country's fuel needs which at the time were being met mostly by wood. One plant on the Ohio River actually produced 30 gallons of shale oil a day over a short period.

In 1859, however, Colonel E. L. Drake drilled his famous crude oil well near Titusville, Pennsylvania, and as was to occur many times in the future, the fresh new supplies of petroleum collapsed the hopes of shale oil entrepreneurs. Compared to the relative ease in drilling into underground pools and pumping the oil to the surface, production of shale oil was an expensive operation involving difficult technological problems beyond the grasp of early oil shalers.

During the years following Colonel Drake's well, the world's leading shale oil producer became Scotland. Scotland possessed good shale deposits and no abundant domestic crude supplies, and an industry developed which operated continuously from 1850 until 1964, when the low prices of imported oil finally forced it to close down. The early Scotch retorting processes, like those in the United States and in France, which had begun producing some shale oil in 1838, were primitive. Most producers used large brick kilns or ovens resembling outdoor barbecues in which batches of crushed shale would be heated externally, usually by coal. After one batch was retorted, the spent shale was emptied and another batch loaded in. The first modern mechanism, developed around 1894 in Scotland, was the Pumpherston Retort. Thirty feet tall, two to three feet in diameter, it contained two significant advances over the batch variety: shale oil production was continuous and the source of heat was, at least in part, the gas and coke by-products. Crushed shale was loaded in the cast-iron top and moved downward by gravity. Heat from burning gas in a brick-lined furnace section moved upward. In the middle, shale oil vapor was formed and was drawn off along with the shale gas. The vapor was condensed and the gas fed back to the furnace. Later modifications employed steam to help sweep the hot gas through the shale and air vents to help burn the residual coke, thus increasing the heat level and the contact of the hot gas with the shale.

The Pumpherston Retort was to serve as the model for most future oil shale industries, but some countries experimented with wholly different retorting ideas. In Estonia, whose deposits supplied fuel gas for nearby Leningrad for many years, a tunnel-kiln was used. Steel cars filled with shale particles rolled along tracks in a tunnel about 8½ feet in diameter and 236 feet long. Heat was injected into the container from a separate fuel source, and shale oil dripped down through perforations in the car bottoms. In Sweden, first tests were made of *in situ,* literally "in place" retorting. The hope was that if ways could be found to retort the shale while still in the ground, the costly problems of mining and disposal of voluminous quantities of spent shale could be eliminated. Holes were drilled into deposits which were about 50 feet deep and quite near the surface of the ground. High voltage electricity was introduced, which heated the surrounding shale and caused shale oil to drip to the bottom of the holes from which it could be pumped to the surface.

As ingenious as many of these methods were, shale oil operations in most countries have been sporadic—between 1875 and 1961 about 275 million barrels of oil were produced, or less than a month's supply for the United States. Most arose because of severe shortages of conventional oil and expired as methods of transporting crude oil worldwide cheaply and efficiently were developed. Too, most foreign shale deposits were too lean and production costs were too high. The only shale oil being commercially produced today is from operations in Estonia and in Manchuria. From the latter the Chinese are said to obtain around 30,000–40,000 barrels/day. A number of American engineering firms and others have held periodic talks with the Brazilian government— that country's shales are second only to those in the United States in richness and Brazil possesses rather scanty crude supplies.*

For many years, people thought the oil shale in the Green River For-

* A 2500-ton-per-day prototype plant is expected to become operational in early 1971.

mation was only slightly more extensive than the Appalachian shale. The only commercial activity was quarrying slabs of shale which, when polished to a lustrous, dark-wood finish, made handsome table tops. This attitude began to change during the early 1900s when, due to the growing numbers of internal combustion engines and the developing European war, concern grew about possible domestic crude oil short-ages. In 1913, the Geological Survey decided to conduct an extensive investigation of the Green River shales which, having been part of a huge 334-million-acre plot obtained from Mexico at the end of the Mexican–American War in 1848 for $15 million, were almost totally government-owned. During the summer of 1915, a USGS geologist named Dean L. Winchester lead a party to survey and sample the lands, and after he returned to Washington in the fall he began computations to determine the total amount of the potentially recoverable shale oil. To his surprise, the figure came to 20 billion barrels, more than three times the estimated national petroleum reserves. Skeptical superiors ordered him to recheck his calculations. The following summer, he returned to the shale lands and after an even more detailed analysis, he reported he had indeed been in error. The figure should have been 40 billion barrels. That year, USGS published Winchester's first comprehensive report, *Oil Shale in Northwestern Colorado and Adjacent Areas,* which concluded that the Green River Formation constituted "a latent petroleum reserve whose possible yield is several times the estimated total remaining supply of petroleum in the United States." In his annual report the following year, Secretary of the Interior Franklin K. Lane stated that "as a result of the investigation of the western oil shales it is believed that it is now commercially feasible to work selected deposits of shale in competition with oil from oil wells, and that these oil-shale reserves can be considered of immediate importance to the oil industry and to the defense of the nation." Newspapers and magazines leaped eagerly upon these ideas. Perhaps the apogee of euphoria was a 1918 story in *The National Geographic Magazine* titled "Billions of Barrels Locked up in Rocks." It gushed:

As the great Creator, through His servants of old, caused water to flow from the rock in the wilderness, so, through twentieth century science, He is causing oil, for ages locked up in the shales of America, to be released for the relief of human necessity. . . .

No man who owns a motor-car will fail to rejoice that the United States Geological Survey is pointing the way to supplies of gasoline which can meet any demand that even his children's children for generations to come may make of them. The horseless vehicle's threatened dethronement [due to oil shortages] has been definitely averted and the uninviting prospect of a motorless age has ceased to be a ghost stalking the vista of the future. . . .

The United States is indeed a country blessed by a generous Providence. . . . No sooner does one of our resources show limitations in production and the pessimists begin to cry, "What shall we do when our reserve is gone?" than immense additional deposits or satisfactory substitutes are discovered. . . . No one may be bold enough to fortell what tremendous figure of production may be reached within the next ten years.

All of this precipitated a wild land rush. Hundreds of prospectors swarmed over the shale lands and under the authority of the 1872 Mining Law staked over 30,000 claims covering some four million acres in Colorado, and perhaps another 120,000 claims in Utah and Wyoming, though totals are difficult to arrive at because many claims were staked on top of one another. Most of the claims in the Piceance Creek Basin were staked along its edges where the Mahogany Ledge outcrops since at the time no one had any knowledge of the rich, deep basin center.

The 1920 Mineral Leasing Act prevented further filing of mining claims for oil shale and said publicly owned shale lands could only be leased by the Secretary of the Interior. But the boom did not abate. More than two hundred oil shale corporations were founded. Many were simply stock promotion outfits which bilked unsuspecting investors on the basis of a few acres of shale holdings and grandiose promises of future bonanzas. A concern called the American Oil Shale and Refining Company set up a small model retort in a booth on State Street in Chicago and sold stock certificates to passersby impressed by the billowing clouds of black smoke which, they were assured, represented pure 100 percent oil, and there was a lot more where that came from.

A large number of the new organizations were legitimate, but their operators found that getting oil from shale was considerably more difficult than panning gold nuggets from a stream or drilling an oil well. Typically, these firms consisted of four or five men who would build a wooden tramway from a cliff-face mine to a crude oven-like retort jerry-built with a variety of belts, pipes, scoops, and the like. A very few, modeled after the Pumpherston Retort, were capable of continuous operation, but most were batch-type contraptions. The presence of scientific controls of any kind was rare. The oil shalers usually found themselves fortunate in the course of a year of sweat, experiment, and expenditure of the group's collective savings to produce a few dozen barrels of oil. The Monarch Shale Oil Company proudly brandished to potential stockholders a certificate signed by the mayor, the head of the Chamber of Commerce, and the chief cashier of the bank in the town of DeBeque, one of several boom towns to spring up on the banks of the Colorado, which read: "To whom it may concern: we personally measured the oil as it came out of the spout [of Monarch's retort] and the flow was continuous and at the rate of 2 gallons each minute." How long this relatively cornucopia-like flow was maintained was not stated, but Monarch's only recorded commercial success was the sale of 25 barrels of shale oil to Glenwood Springs, a town near Rifle, to oil the streets. The output of most retorts ended up as such somewhat ignominious products as sheep dip, soap, dandruff treatments, tars, and various medicinal remedies.

The most ambitious shale undertaking of this era by far was Catlin Shale Products Company, organized by Robert M. Catlin, a native of Burlington, Vermont. In 1875, while working as a mining engineer, he became intrigued by smouldering fires in dumps of waste material near Elko, Nevada, that had been left by the Central Pacific Railroad Company while making tests in coal beds along its right-of-way. He discovered the source of the fires to be beds of paraffin shales, which are similar to oil shales except the organic content of the latter is higher. He resolved to discover whether a commercial operation was feasible. A member of the American Institute of Mining and Metallurgical Engineers and former chief mining officer with the New Jersey Zinc Company, Catlin had no illusions about what he was attempting. "One of the unfortunate features of the matter," he wrote to a friend, "seems to me is that anybody can produce oil from the shale so easily. People take a crooked pipe and in an ordinary fire get some oily results and the imagination does the rest."

Catlin purchased and leased some land from Central Pacific in the late 1890s and over the next thirty years invested perhaps as much as a half million dollars of his own money—he never sold stock—in a series of three retorts. The last, built in 1921, was quite sophisticated: continuously operating, internally heated, thermally self-sufficient, it was 40 feet high, and assisted by a 16-man crew, it could process 250 tons of rock daily and produce nearly 50 barrels of shale oil. Uniquely, the company was fully integrated: it mined rock, crushed it, retorted it, refined the oil, and even retailed it through a rudimentary marketing organization. Though Catlin Shale Products did not earn a profit as such, it did find a market for many of its products. Fuel oil was sold locally to the power company and to individuals for heating and powering small tractors and motors. Lubricating oil was sold through garages in Nevada and California. Several hundred tons of paraffin wax were shipped overseas—indeed paraffin wax was used by New Jersey Zinc in zinc production, which was probably a principal reason Catlin persisted with the more paraffinic Nevada shales. Union Oil Company of California, the most active of the major oil companies in shale during this period, made a thorough study of his process, though it eventually took no action.

For reasons that are not entirely clear, Catlin closed his retort down in 1924. He apparently had regarded his operations until then as mainly an experiment and had felt that to run his company on a profitable basis he would have to raise a large amount of capital to improve substantially the capacity of his equipment. Especially he needed money for an improved refinery, since many of his products were inadequately distilled and purged of impurities and had difficulty competing against the products of crude oil refineries. The problems he encountered raising funds are unknown. One was probably a decline in crude oil prices due to new oil field discoveries in California, Texas, and Oklahoma. By 1924 the price of crude had dropped to less than half its 1920 level, and prices were to decline much more before the end of the decade. Further, the lean, thin (less than five feet), often tilted and folded shale beds near Elko probably seemed to potential investors a far less likely location for a shale oil industry than the Piceance Creek Basin. Catlin himself was by now ell into his 70s, and his zeal may have been waning. In any event, on December 23, 1930, Catlin Shale Products Company, the United States' first and only continuously operating commercial shale corporation, was formally dissolved. Catlin assigned remaining assets, less than $3,800, to his son.

By that year, most of the other oil shale corporations had long since either gone bankrupt or had been otherwise dissolved. Those that had not were soon vanquished by "generous Providence" which in 1930 blessed the nation with the gigantic six-billion-barrel East Texas oil field—until the discovery of the Prudhoe Bay fields on the North Slope of Alaska, the largest ever found in North America. This caused such a glut that crude prices dropped as low as 10¢ a barrel. (Crude prices today average around $3.10 a barrel at the well.)

To this day, the oil shale industry has never revived. There have been periods of heightened interest, caused by fears of crude oil shortages, and at one point students at Cornell University spent six years drawing up plans for a city of 300,000 they felt certain would soon rise up near Rifle. But the only commercially operating shale industry today consists of a man named Mike Cross, a local Rifle jeweler who sells polished oil shale pins and tie clasps to tourists.

The occasional articles on oil shale that have appeared in magazines and newspapers over the past few decades—typical titles are "Colorado's Fabulous Mountains of Oil," "Is Shale-Oil Boom on the Way at Last?" and "Giant Oil Shale Industry is Waking"—have had a common theme: there is certainly a great deal of oil shale out there in the Colorado mountains, but unfortunately it just costs too much money to get the oil out of the shale. "Why have Colorado's immense oil shale deposits gone unexploited for so long?" asked a *Reader's Digest* story in 1957. "It is a story," the magazine replied, "of baffled men against rugged mountains." These stories all add that large numbers of people are laboring assiduously to lick the technological problems, and it will likely be only a matter of time before the rugged mountains begin to yield their vast treasure.

Belief has been solidified, therefore, in the minds of the public, most government officials, and even some people with broad acquaintance with oil shale, in two apparent truisms:

1. *The reason there has been no development of the oil shale is because development is not yet economically feasible.*

2. *As soon as it becomes economically feasible, oil shale development will occur.*

It does not require much investigation to discover serious flaws in these "truisms." One reads for instance numerous statements that shale oil production is "nearly competitive" with crude oil production. Why then, especially if the oil shale deposits are so rich, is hardly any research being directed toward developing shale oil production techniques?

Attempting to get beyond the truisms is like wandering through a massive, high-walled labyrinth enveloped in smoke, with many false exits and dead ends. Finally, two large controlling entities are revealed: the federal government and the oil industry. The federal government, specifically the Department of the Interior, as the guardian of the 85 percent share of the shale contained in public land, has always been considered the logical party to encourage development of the shale land. The oil industry, as the possessor of petroleum technology and the refining and marketing organization through which shale oil must eventually flow, has always been considered the logical party to develop the shale. Both, though, have acted not unlike a reluctant bride and bride-groom at a shotgun wedding. In the face of demands for action, they have inched forward slowly, ever so slowly, toward the altar, but their hearts have not been in it. At the rate they have been progressing, they will not get there for a long, long time. Meanwhile, the old oil shalers in the dusty towns on the banks of the Colorado continue to gaze long-ingly at the white cliffs and wait and hope.

PART TWO

"Many times in the history of the United States technological change and new discoveries have created new industries while partially or completely destroying others. Recent evidence suggests that another case of creative destruction is in the offing, with shale oil production supplementing, and in time partly displacing, the long-established crude petroleum industry."

—Dr. Henry Steele,
Professor of Economics,
University of Houston,
1963

"I think, to make the principle here, Senator, what would be important to me is that I would like to have a little assurance about what the running rules of the whole game are going to be a little bit further down the road . . . to have them nailed down, so you are sure of what environment you are going to have to operate in."

—Charles F. Jones,
President,
Humble Oil & Refining Co.,
September 14, 1967
(explaining to the Senate Interior
Committee his company's reluctance
to begin commercial shale oil
production)

Chapter 2:
The oil industry,
which one would think
ought logically to be developing the shale,
is very apprehensive,
and for good reason.
For one thing, their established crude oil business
could be seriously disrupted.

Reading the speeches and public statements of oil industry executives on oil shale, one gets the feeling that trying to figure out ways of economically producing shale oil is like chasing a will-o'-the-wisp such as an end to poverty. Despite our strivings, it will continue to elude us until the arrival of some distant millennium. A recurring metaphor compares oil from shale to gold from the sea, as in the following classic articulation by C. E. Reistle, Jr., chairman of Humble Oil & Refining Company, United States subsidiary of Standard Oil Company (New Jersey), in a speech in March, 1966:

. . . No practical oil man will be misled by what may seem to be a mammoth supply of oil in the shale, just ready for the taking. There is a big difference between this "on the books" oil from shale and oil in a pipeline. It is the same difference that exists between gold in the sea and gold in Fort Knox. It has been estimated that the oceans of the world contain more than six million tons of gold. That is enough to make a pile which would weigh almost as much as the Great Pyramid of Egypt. Figured at today's market value, that pile of gold would be worth *seven trillion dollars*. By contrast, the present monetary gold supply of the entire world is worth only *forty billion* dollars. But you don't see anyone rushing to mine the sea for gold. And no wonder! A ton of sea water contains only 2 to 60 milligrams of gold— not worth the effort with our present knowledge.*

* This analogy is somewhat misleading. There are more dollars worth of oil in 1,400 square miles of land containing perhaps 1,000 cubic miles of rock than there are dollars worth of gold in 140,000,000 square miles of ocean containing 317,000,000 cubic miles of water.

The plain facts are, it will be necessary to spend large sums on research and development before a profitable shale oil industry can become a reality. Substantial progress is being made in oil shale research. Technology has now progressed to the stage where we may hope to produce and process oil from shale at a cost closely competitive with crude oil. But we still have a lot to do if we are going to get those costs closer together. . . .

Let me emphasize that I do not mean to imply that oil and gasoline will not eventually enter the market. Today, few question that. My company would not be spending money on research unless we were reasonably inclined to believe that shale oil can one day supplement other energy sources. Nor would our competitors! But the old question of economics remains a deciding factor. . . .

The implication of Reistle's speech is that the oil industry is working zealously on shale oil technology, and that as soon as production of shale oil becomes cheap enough, oil men will applaud vociferously and joyfully begin building commercial plants.

The available evidence indicates otherwise.

Scores of interviews with oil executives, government officials, economists, and other experts, as well as published industry and government studies, testimony before Congressional committees, and documents filed during recent litigation, lead to four inescapable conclusions:

1) The major activity by far related to oil shale pursued by the major oil companies to date has been the acquisition of oil shale land. According to testimony in 1967 before the Senate Subcommittee on Antitrust and Monopoly by Interior Department Deputy Solicitor Edward Weinberg, 13 large oil companies, including the nation's seven largest, own interests in some 228,200 acres of shale land out of the 349,000 acres currently in private hands. Among the largest owners were Standard Oil Company of California with 40,309 acres, Union Oil Company of California with 32,870 acres, Mobil Oil Corporation with 24,-192 acres, Getty Oil Company with 23,498 acres, Texaco Inc. with 14,401 acres, and Cities Service Company with 12,180 acres. Documents subpoenaed recently from a number of large oil companies by the Bureau of Land Management, a division of the Interior Department, for Colorado Contests #359 and #360 contesting the validity of some pre-1920 oil shale mining claims indicate this figure is low, and that it does not include a sizable number of options to purchase, various partial interests in land listed in the name of private individuals and substantial ownership of land claimed under the mining laws but for which actual title is still held by the government. Exact figures are impossible to ob-

tain, but it would appear that these land holdings represent an initial investment by their purchasers in excess of $50 million and perhaps considerably more. Exercise of options would raise the total significantly. Shell, for instance, holds an option to purchase a 21,120-acre block of land for over $42 million. The total potentially recoverable oil contained in these private lands is at least 300 billion barrels.

Oil companies have spent additional sums acquiring water rights in the shale land area. Many oil shale experts, though by no means all, believe that a shale oil industry will require vast quantities of water, not only for actual shale oil production but for normal use of the industry's employees. Present water supplies, however, are quite limited. According to documents submitted to the Antitrust Subcommittee by Edward Weinberg, most of the major oil industry shale landowners have either purchased sizable acreage with appurtenant water rights or have made filings to the Colorado River Water Conservation Board. Humble Oil reportedly spent over $1 million between 1966 and 1968 obtaining water rights by means of what some residents claim was extremely active lobbying with members of state and local governments.

These moves, along with its land ownership, have given the oil industry a kind of de facto domination of the area's immediate potential for commercial shale oil production. Northcutt Ely, a former Interior official and now a Washington lawyer who has been acquainted with oil shale since the 1920s, told the 1967 Oil Shale Symposium conducted by the Colorado School of Mines that: "The sophisticated oil companies have bought up water from agriculture from the State, from the BUREC [Bureau of Reclamation] and have bought reservoir sites. If it takes 1.2 barrels of H_2O to produce a barrel of [shale] oil, it can be accurately predicted today who controls the oil production."

2) In distinct contrast, with two or three significant exceptions, expenditures on oil shale research and development by major oil companies have been negligible. Most oil companies keep details of their research budgets confidential, but evidence submitted for the BLM contests produced specific figures from four oil companies, all of whom are regarded as having above average interest in oil shale. Between 1950 and 1966, Shell Oil spent $1.5 million, an average of $88,000 a year, on oil shale research, which amounted to 1.6 percent of the company's total "research and exploration" expenditures. (To maintain its option on the block of shale land, Shell is obligated to pay a maximum of $280,000 annually.) Since 1964, Mobil Oil has been spending about one percent of its annual research budget on oil shale for a total of

about $1.2 million. Mobil meanwhile has spent over $3 million on shale land and has an option to pick up further acreage for $1.2 million. Between 1962 and 1967, Texaco spent an annual average of under $100,000 on shale research or .3 percent of its $29 million research budget for 1965. Union Oil of California spent $64,108 or one percent of its research outlays on oil shale in 1966.

Other companies have done even less. During hearings for the BLM contests, David M. Anthony, an executive with Cities Service, responded as follows to questions from an Interior attorney:

Q: Has the company done any work since 1950 involving the mining of oil shale in the Green River Formation?
A: No, sir.
Q: Has the company done any work since 1950 involving the retorting of the oil shale of the Green River Formation?
A: No, sir.
Q: Has the company done any work since 1950 involving the refining of shale oil?
A: No, sir.

For this book, a survey was made of the oil shale activities of 20 major oil companies. Executives from the following nine companies responded that their company had no research program on oil shale: Getty Oil, Skelly Oil Company, Continental Oil Company, Standard Oil Company (Indiana), Phillips Petroleum Company, Marathon Oil Company, Amerada Hess Corporation, Sun Oil Company, Kerr-McGee Corporation. Executives from nine other firms said their research programs on shale were, variously, "small," "very modest" or "not significant": Cities Service, Standard Oil of New Jersey, Standard Oil of California, Texaco, Shell, Union Oil of California, Occidental Petroleum Corporation, Gulf Oil Corporation, Mobil Oil. The only companies who said they were spending any significant sums on oil shale research were Atlantic Richfield Company and Standard Oil Company (Ohio). These two very special exceptions will be discussed in detail in Chapter Four.

3) The research programs conducted by oil companies have consisted primarily of laboratory work, accompanied by the filing of periodic patents, and various "paper" economic studies based on published figures from the Bureau of Mines and other mainly government sources. During the past 40 years, only a small handful of efforts has been made to construct actual experimental shale oil facilities. Union Oil of California has been the only company to build a pilot plant—this achievement

will be discussed in Chapter Three. Since then, no oil company has attempted to duplicate the feat. Nobody since Catlin Shale Products has ever tried to build a commercial-scale shale plant.

4) The philosophy behind the oil industry's oil shale activities has been to acquire a hedge against the future. A 1959 Standard Oil of California memorandum filed with the BLM Contest evidence states: "The company's long term objective on oil shale is to secure and maintain reserves of satisfactory quantity, quality, and location as insurance against possible future crude oil shortages or changes in economics for shale as a competitive supplementary source of oil." Texaco's 1966 Annual Report in discussing its shale land holdings said Texaco "will be in a favorable position to produce shale oil when it becomes attractive to do so."

Similar ideas lie behind research efforts. A February, 1961, report by E. W. Clark, special assistant to the vice-president of Shell Oil, recommended that the company

take a somewhat more active interest in current research and new developments in the thermal treatment of [oil shale] formations underground. . . . The main objective of these modest measures would be to protect Shell's competitive position against the possibility that current research into *in situ* thermal recovery techniques may result in the evolution of substantially less costly methods of shale-oil production and thus bring commercial development of oil shales much nearer to realization.

The BLM evidence contains numerous oil industry memoranda, reports, and other documents containing phrases such as: "Keep in touch with industry and government developments," "Keep abreast of competitor activities," and "Follow activities described in the journals."

These phrases were also prevalent during interviews with the oil executives questioned for this book, most of whom would only permit quotation without attribution to them by name. Getty Oil's attitude, said an executive, was "wait and see." Cities Service: "Our effort is mainly defensive. If anything comes along that our company might miss out on, we look at it." Marathon: "We're keeping up with the trends." Amerada Hess: "We're doing just a casual keeping up with developments. Anything that we would do would be defensive." Standard Oil of California: "Some ideas pop up and we do research on them. We're not going to stop thinking about it. We're always looking. But we have no full-time research staff." Occidental Petroleum: "We're not doing anything other than keeping abreast of what others are doing to see how we might get involved." Mobil Oil: "We want to keep our options open as long as possible without spending a lot of money."

The apparent assumption that would appear to underlie these statements and activities is that the economic attractiveness or lack of attractiveness of commercial shale oil production has somehow been incontestably determined; that at the moment it costs precisely x number of dollars to produce a barrel of shale oil in a commercial-sized plant; that this figure is unquestionably too high and so irreducible short of a technological miracle that pursuing the matter further is senseless.

Again, the available evidence indicates otherwise.

Precise costs, in fact, have never been accurately established, and most of the best available guesses are extrapolations of data over 15 years old. *The Oil Shale Policy Problem,* a synopsis prepared in 1964 by the Interior Department for the Oil Shale Advisory Board, a group of private individuals appointed by Interior Secretary Stewart Udall to advise him on oil shale policy, states:

As far as is known, no comprehensive, completely original studies of the economics of an oil-shale industry have been made since the National Petroleum Council [an advisory group of oil and gas industry executives] and the Bureau of Mines evaluated the situation in 1951. The results of those studies were in good agreement, and each showed that a shale-oil industry at that time would be nearly competitive with the petroleum industry. The values calculated in 1951 are now obviously far out of date. . . .

Since 1951—those studies will be taken up in more detail in Chapter Three—numerous factors have acted to increase costs, such as labor, materials, and the like. Others, such as technological developments, have undoubtedly lowered costs. Still others are difficult to assess, such as the complex changing patterns of the whole energy industry.

More recent inputs for paper economic studies have been limited field tests, usually sampling of core drills, and modest laboratory experimentation with "bench"-scale apparatus. Most mining engineers contend that figures derived from such work are rather dubious, and that the only way to obtain reasonably accurate cost figures for something as unproven as commercial extraction of shale oil is at least to build and operate a sizable pilot plant. "All of these slide rule numbers are fictitious—You really don't know until you put something on stream," says Irvin Nielsen, a Denver geological engineer who has been making investigations of the shale lands for many years. "If you happen to have an ingenious engineer, you might find yourself with a big profit. If your man isn't ingenious, you might lose money. Even after your plant is working, you might figure out that if you up the pressure in your retort a bit, you'll get three or four times as much throughput, and all of a sudden you're 50 percent more profitable. You just can't know these things accurately beforehand." Dr. Henry Steele, Professor of Economics with the University of Houston said in an October, 1968, article in the *Natural Resources Journal* on oil shale: ". . . (I)n view of the crucial importance of the level of production costs to the entire shale oil debate, it is remarkable that so little insistence has been placed on the presentation of comprehensive and authoritative production cost estimates."

Oil executives do not deny the possible inaccuracy of their estimates. In a letter in 1967 to Senator Henry M. Jackson, chairman of the Senate Interior and Insular Affairs Committee, Humble Oil President Charles F. Jones wrote: ". . . (T)he state of the art for producing oil from shale is [in] an embryonic stage, and the costs are difficult to predict." A Shell report on *Current Prospects for a Shale Oil Industry,* quoted in the BLM evidence, said that ". . . there are many divergent and conflicting opinions concerning the technology and economics of producing the [shale] oil."

Numerous recent studies, nevertheless, have predicted that shale oil production would be commercially attractive. Many of the parties making these studies have a financial interest in shale development, though since their interest is presumably based on assumptions of shale's attractiveness, the validity of their figures should not necessarily be immediately discounted. One of shale's biggest boosters has been The Oil Shale Corporation, known as "Tosco," whose activities will be described in detail in Chapter Four. A small company whose backers are generally unrelated to the domestic petroleum industry, Tosco has been the only party actually to build and operate a shale oil production facility of any size during the past 15 years. In July, 1969, Hein I. Koolsbergen, Tosco's president, said his company's technology would be able to produce a barrel of shale oil commercially for an operating cost of $1.25 a barrel, not counting land acquisition costs but including a credit for production of various by-products. Adding a 50¢ per barrel transportation cost to a West Coast refinery, a barrel of shale oil could be delivered for $1.75, about $1.50 less than the selling price in California for conventional crude. Previously Koolsbergen had estimated a commercial plant using Tosco's technology could yield a return between 15 and 20 percent measured on a "discounted cash flow" basis, which is similar to return on average invested capital. (A 15 percent return is currently generally accepted as the minimum acceptable to investors for a new risk venture such as shale oil production.) Koolsbergen concluded at the 1969 meeting that production of shale oil "is a very attractive and very profitable business" though Tosco had not yet actually put a commercial plant into operation. Dr. Charles H. Prien of the Denver Research Institute agrees with the Tosco figures, which he helped develop, and states that shale oil is "competitive with current petroleum costs."

Russell J. Cameron, head of Cameron Engineers, a Denver consulting firm with wide experience in synthetic fuels, says: "With the processes we now have, shale oil can definitely compete in today's market, and with an adequate return on investment." At the 1968 meeting of the American Association of Petroleum Geologists, Cameron noted that there were perhaps 300 square miles in the center of the Piceance Basin where some 300 billion barrels of oil were contained in shale with yields up to 25 gallons per ton. "The almost unnoticed breakthrough in materials handling that has taken place in this past 20 years," he said, would permit recovery of the shale in the basin center by means of a huge open pit mine employing giant shovels with perhaps 500 cubic yard bucket capacities which would vastly reduce mining costs. "From such a reserve," he said, "we could produce at least 10 million barrels of shale oil per day for a hundred years with additional production from lower grade shales and thinner beds of rich shale." Such shale oil, he predicted, "will be competitive with oil produced anywhere."

These estimates are not at wide variance with two of the most comprehensive recent studies by more disinterested experts. At the 1967 oil shale hearings held by the Senate Antitrust Subcommittee, Dr. Henry Steele, of the University of Houston, who is a respected petroleum expert, described the results of his research over a ten-year period on shale oil production costs. "It is my view," he told the committee, "that in all likelihood shale oil costs are at present low enough to allow crude shale oil to be produced profitably at current crude petroleum price levels." He estimated that, based on 1965 figures, the total cost of a barrel of shale oil from a 25,000 barrel/day plant shipped by pipeline to Los Angeles would be $1.96, including a 48¢ pipeline charge. The margin between this and the then current West Coast crude price would yield, he said, a 14.7 percent after-tax return. "The cost summary given," he said, "is conservative not only in that the intention is to overestimate costs rather than underestimate them, but it also tends to underestimate by-product revenues." He did not include alumina, sodium compounds, and other minerals which some geologists, as Chapter Seven describes, feel may be produced in quantity along with the shale and would perhaps be worth as much as 50¢ a barrel or more. In his article in the October, 1968, *Natural Resources Journal,* Steele updated the previous figures, by adjusting for inflation and technological advances and including figures for a 250,000 barrel/day operation, which would result in large scale economies of pipeline transportation. The return on average invested capital, after taxes, for a 25,000 barrel/day plant, he said, would be 16.5 percent. For a 250,000 barrel/day plant, it would rise to 22.2 percent.

The most thorough recent government study on shale oil costs was made in the Interior Department's *Prospects for Oil Shale Development* published in May, 1968. Assuming construction of a 60,000 ton/day shale plant costing $138 million, it made a series of cost and return estimates depending on various stages of improving technology. If construction were begun in 1970 and did not hit any serious snags, the plant in 1972 could begin to sell shale oil for $2.98 a barrel, exclusive of transportation and land acquisition charges, and earn a 12 percent discounted cash flow return. (This figure includes profit, while Steele's and Tosco's per barrel estimates did not.) Operating costs, said the report, would be $1.49 a barrel, but by 1976, with an "improved first generation plant," they would decline to 97¢ a barrel, meaning the plant could sell shale oil for only $2.12 at the retort and still earn a 12 percent DCF return. By 1980, the report continued, a "second generation" plant could be on stream embodying significantly improved retorting and mining techniques, and could generate a 12 percent DCF return with a selling price of $1.58, and a 20 percent DCF return with $2.78. Presumably these costs could continue to drop as technology improved and economies of scale were exploited.

Economic studies by the oil industry on shale oil costs, as exhibited at Senate hearings and in the BLM evidence, come to a range of conclusions, but there is a general consensus that shale oil could be produced right now at a profit. Often the degree of profitability is estimated to be close to or even better than what the oil industry achieves on its present operations. During the Senate Interior Committee's 1967 hearings on oil shale, Humble Oil President Charles F. Jones stated that according to "our current engineering studies . . . it appears that an oil shale industry might be developed on the basis of reasonably anticipated technology in the short-range future, such as the next decade, that would be competitive with the average [return] of mining and manufacturing, which is in the neighborhood of 12 percent or a little better." In 1968, Standard Oil of New Jersey earned 12.3 percent on its net worth.

Some of the most detailed recent paper economic studies have been made by Shell Oil, whose records and files are represented in the BLM evidence in great detail since Shell owns some of the oil shale claims the government is challenging. In 1964, Arthur Brown, Shell's Denver manager of production, attempted to convince his superiors in New York of a "five-year plan for the development of a commercial process for the production of hydrocarbons from oil shale." "The ultimate goal of our program," Brown wrote, "is to place Shell in a position to produce shale oil in the early 1970s, if it chooses to do so." Between 1964 and 1967 several financial analyses were made, all of which stressed the sizable probabilities of error. One that analyzed the property on which Shell possesses its $42 million purchase option—obtained from a man named Tell Ertl—stated:

Retort costs used in Shell's preliminary economic feasibility studies were obtained from limited information published by the USBM[U.S. Bureau of Mines]. . . . The degree of accuracy of the costs employed in our Ertl property profitability analyses for retorting rich shale (underground mining) is not known, and the costs for retorting low grade shale (open pit mining) are subject to even more question than those for retorting rich shale.

Nevertheless, the studies predicted returns from a large commercial shale oil plant ranging from 5.8 percent to 22 percent, depending on the various technological assumptions. In 1966, however, Brown's superiors decided against pursuing the program, primarily, they told Brown, "since we have very little acreage where oil shale mining might be undertaken, and it is questionable whether we will acquire more of this type acreage." However, when R. G. Christianson, Shell's executive vice-president for exploration and production, was asked during the BLM hearings whether he knew "anything that would prevent Shell Oil from going out and buying patented oil shale claims" [claims for which the government has relinquished title] he replied, "Not necessarily, except the simple state of the art as it is known today and the competitive yield that one might get from this sort of an activity as measured with other opportunities that one chooses between them." The BLM evidence quotes T. W. Nelson, Mobil's senior vice-president for exploration and production, as saying in 1964 that ". . . (W)e do not feel that there is a lack of shale reserves in private hands so as to impede the development of an oil shale industry." Shell's optioned Ertl block is one of the most sizable well-situated single pieces of shale acreage in private hands, and Arthur Brown testified at the BLM hearings that the plot "could be worked at a profit that would be comparable with the rate of return that we get from our current exploration and production ventures." Shell in 1968 earned 12.3 percent on its net worth.

A number of other oil companies contend that the estimated degree of profitability is too low to lure them into something which, as Charles Jones of Humble told the Interior Committee, "obviously has hazards the traditional industry does not have as many of." T. W. Nelson of Mobil stated in a deposition for the BLM contests that "it is my opinion that the remaining technical uncertainties can be resolved on a basis that, with economics as we see them, would bring a positive rate of return. This is to be distinguished from a satisfactory return."

In short, oil industry executives generally agree that producing shale oil is profitable and probably "closely competitive" with crude oil. But they say it isn't close enough. They have no firm idea of what the real production costs or profits might be because they have never wanted to spend the money for a commercial plant or even to conduct anything more than the barest amount of protective research. Meanwhile they have invested many millions of dollars in extensive tracts of oil shale land.

Seemingly, it is a bit of a paradox, leading to some obvious questions. Why aren't oil companies doing more research? Why aren't they building pilot plants to help discover the real costs? Why aren't they eagerly trying to develop a resource—or at least to find out whether its development would be worthwhile—that would solve their own oil needs, and those of the country, for decades to come?

Oil executives become somewhat uneasy when this line of questioning is pursued in interviews, and they often attempt to quell it by asserting they really are doing a lot of research or that they are very enthusiastic about shale, sometimes to the point of contradicting statements made earlier in the interview that their research programs were very modest.

Illustrative of this attitude is another group of Shell Oil documents in the BLM evidence relating to Shell's decision in 1950 to begin acquisitions of shale land. A letter in January by the Pacific Coast territory vice-president said:

During the discussions with Shell Development several possible reasons which might be regarded as being for the acquisition of these lands were examined. These arguments stem primarily from the publicity which synthetic fuels have received in recent months and pose the question as to whether either the financial people, generally, or the Government would look critically upon the lack of participation on the part of a major company, such as ours.

In October, a Mr. Burns, then Shell's president, decided to go ahead. A memorandum summarized his action this way:

Mr. Burns particularly stressed a point which had hitherto not been mentioned in correspondence, namely, that it would be of considerable value to Shell Oil vis-à-vis financial and political quarters to be able to state that Shell Oil was not only carrying out intensive research on the processing of shale oil, but was also in possession of large potential reserves thereof.

Shell's Annual Report that year stated: "Progress in the production and refining of shale oil is being followed closely. While the ultimate economic significance of these investigations cannot be appraised at present, they are essential in order to protect your company's position in an industry characterized by constant progress and new developments."

The fact that these investigations turned out to be rather diminutive prompted a letter in 1955 by the company's Los Angeles exploration manager which criticized the

. . . evident lack of interest on the part of the Company in shale oil as a secondary source of refining raw materials, this despite the fact that shale oil comprises, in one relatively concentrated area, the largest oil reserve in the U.S.—more than 125 billion barrels. Shell Development Company has not conducted any research on extraction or refining of shale oil and none is programmed for 1955. . . . This attitude appears to stem from Economic Development's evaluation which seems to indicate a 6.5% rate of return for an oil shale operation. . . . Our comment: in view of the size of the oil shale reserve, however, and the certainty of its ultimate exploitation, it would seem that such an evaluation of such a futuristic project is scant justification for not conducting basic research, which conceivably could reverse our entire outlook. . . .

The common rejoinder of many oil industry executives to suggestions they may be less than enthusiastic about shale is a statement made by Charles F. Jones of Humble to the Senate Antitrust Subcommittee in 1967. "I would hazard a guess," he said, "that in recent years private industry has spent $100 million" on oil shale. $100 million sounds like an ambitious, wholly respectable figure until it is broken apart. Jones indicated he included land acquisition costs, which total at least $50 million and become much higher when items like taxes, interest, option maintenance payments, and other continuing charges are added. Standard Oil of Ohio alone spent or obligated itself to spend $28.8 million on shale land between 1964 and 1967. Nearly $28 million had been spent by 1967 by The Oil Shale Corporation, which is a special corporation organized specifically to develop oil shale and which is not part of the conventional oil industry. Some $12 million was spent by Union Oil of California in an abortive research effort abandoned in 1958. All told, it is doubtful that, with the exception of the special case of Atlantic Richfield (described in Chapter Four), the oil industry is spending much more than a few million dollars annually on oil shale research. In contrast worldwide capital spending by the industry totals over $15 billion a year, including close to $5 billion for exploration and development of new domestic petroleum reserves. Total research outlays in 1967, according to the American Petroleum Institute, were $469 million.

The industry's spending on shale is put into perspective by its activities in response to the discovery of huge oil fields on Alaska's North Slope by Atlantic Richfield and Humble Oil. In September, 1969, the industry bid $900 million for leases in the area from the Alaskan government, and one company, Amerada Hess, bid $83 million, just $6 million less than its 1968 net profits. To get the oil out, a consortium of Atlantic Richfield, Humble, and British Petroleum and five other companies announced plans for an 800-mile, 48-inch pipeline across Alaska which would cost at least $900 million (now estimated at $1.7 billion) and would be the world's largest private construction project. The consortium said it was also studying plans for an additional 2,600-mile pipeline from Puget Sound in Washington across the country to the East Coast. Total cost of a pipeline from the North Slope to Chicago to the East Coast has been put at $4–$5 billion. Meanwhile Humble appropriated $44 million to determine whether it would be feasible to bring oil out by tanker through the Northwest Passage. Even if it is, a Humble official estimated it would cost between $150 million and $500 million to build docking and loading facilities in Prudhoe Bay. About 30 new ice-breaking tankers would have to be built at a cost of at least $1.5 billion. Frank N. Ikard, head of the American Petroleum Institute, said the oil industry would spend $5 billion during the next five years "with relatively little return in that period." According to one estimate, $300 million was to have been spent by the end of 1969 alone, not counting the lease bids. Some experts place total capital investment at $10 billion.

Consider, now, the "hazards" of the North Slope. Despite all the planned and actual spending, the oil industry asserts it has only a vague idea how much money it will make out of Prudhoe Bay, and it is not an absolute certainty that much money will be made at all. The eventual costs and problems of drilling oil from some of the world's most forbidding terrain can only be guessed at. Even after the icebreaker *Manhattan* managed to crash its way through the ice in the Northwest Passage to Prudhoe Bay, Humble officials said it would take at least a year to evaluate the trip. Another trip began in April, 1970. Constructing the docking facilities, according to a *New York Times* reporter, will be a "monumental challenge" due to thick ice, shallow water, and powerful tides, and he quoted some observers who doubted it could be done at all. The pipeline consortium is in no better shape since they can only guess at the problems of shipping hot oil (it comes out of the ground at around 160 degrees F.) across mountainous tundra covered with layers of permafrost. Regulations imposed by the government to protect the Alaskan wilderness against massive spillage caused by pipeline rupture as well as other possible desecration are bound to be stringent. Legal and political challenges to the pipeline's construction that developed during the spring of 1970 were estimated to have delayed the start of construction until the middle of 1971. And on top of everything else, no one is sure at all just how much oil there is on the North Slope nor how much of this is economically recoverable. Estimates of the total reserve range from 5 billion to 50 billion barrels.

Predictions of the ultimate profitability of the North Slope range all the way from mammoth—by a Cabinet task force appointed by President Nixon—to modest—by the oil industry. Both groups, of course, may well have special reasons for their conclusions: the task force is interested in removing import controls, which would lower the price of domestic crude oil, while the oil industry is always sensitive to suggestions it might be making excessive profits. Atlantic Richfield estimated the delivered price to the West Coast would have to be very close to the current price of around $3.10 while the task force said it would be only $1.11. The consensus of disinterested opinion seems to be that the eventual figure would be closer to Atlantic Richfield's guess.

In any event, in refuting the task force report Humble Oil pointed out that despite a $1.9 billion investment in Alaska's Cook Inlet oil field, the oil industry is still $1.5 billion in the red on the project, and that the eventual return on total investment would only be between 3 and 6 percent. Due to high lease costs from intense competitive bidding, investment in Louisiana offshore land has returned only 4 to 7 percent. Despite a $603 million investment in Santa Barbara Channel bids in 1968, the industry at the beginning of 1970 had not yet received permission to resume full-scale drilling due to continuing controversy over the huge oil-spill crisis in January, 1969. Shell noted that the industry was still some $7.5 billion in the hole on its $16 billion investment in all offshore oil operations. Yet in contrast to its $1.5 million expenditures on oil shale research over 17 years, Shell bid $1.2 billion for offshore tracks off Texas and in the Santa Barbara Channel in 1968. These bids are true "off the top" expenses in that they are over and above actual development and operating costs and are unrecoverable even if a tract should not yield a single barrel of oil.

The hazards of starting an oil shale plant are not recommended for faint hearts either. Most estimates indicate initial capital expenditures to develop the technology and construct a plant to produce 100,000 barrels a day—perhaps 10 percent or less of the requirements of the largest oil companies—would run to several hundred million dollars. A million barrels a day of productive capacity might cost $2 billion for production facilities alone. Though quite firm costs could be obtained from a much less expensive pilot plant, there would be, as in Alaska, no guarantee that unforeseen hitches would not develop to balloon costs of a full-scale commercial plant. Such problems as waste disposal are very sizable.

Still the risks do not seem excessive. Edward F. Morrill, senior vice-president of Standard Oil of Ohio, which has been participating in a research program with The Oil Shale Corporation, testified at the BLM hearings as follows:

Q: Is it correct, Mr. Morrill, that in your opinion, once a decision is made to move to a commercial oil shale operation, that it will be an assured successful operation?

A: Outside of death and taxes, I don't know of anything that is assured. But I would say that the unique part of this business project which is not normal to a new product development is having an assured market, for example. So having an assured market coupled with a good development program, I would say that this project, when it reaches the point of commercialization, should have few risks that are in addition to the normal business risks of any new project. . . . I am now answering that in the frame of reference of the same business risks we would have in drilling a development oil well, building a refinery. That is literally what we are doing.

In comparing the relative risks between the North Slope and oil shale, one must also consider the rewards. A recent Chase Manhattan Bank study said the North Slope could eventually probably be made to produce oil at the rate of a million and a half barrels daily, less than 8 percent of the country's daily oil needs of 18 million barrels in 1980. How long this output would last depends of course on the total recoverable reserves. At the moment, estimates above 25 billion barrels appear to be somewhat fanciful. Only two oil fields containing more than 5 billion barrels have ever been found outside the Middle East, one in Venezuela and the other in East Texas; the latter is the largest ever found in 110 years of exploration in North America. The general consensus is that the North Slope probably contains recoverable reserves somewhere between 10 and 15 billion barrels, which at maximum would constitute in aggregate a three-year supply for the United States.

The oil shale reserves, including the land owned by the government, are virtually inexhaustible. A recent Bureau of Mines study said eventually 6 million barrels a day could be produced from the Piceance Creek Basin, and other estimates have run to 10 million barrels and above. In his "gold in the seawater" speech, C. E. Reistle, Jr., of Humble did note that just a 50,000 barrel/day shale plant, which he estimated would cost $100 million, "would be comparable to finding, developing, and producing a 365 million barrel oil field—and we don't find many fields that size. In fact, there have been only 23 fields which have produced that much oil in the history of the domestic petroleum industry."

Clearly, there must be something besides Charles Jones's "hazards" and Reistle's "old question of economics" that is deterring the oil industry.

A strong deterrent, one discovers, is the very alien nature of shale oil production to most oil men. Producing crude oil involves finding oil fields, drilling wells, and getting the oil to the surface of the ground. Producing shale oil is essentially a mining operation—the only significant difference between producing copper and shale oil is the end product. The chief technological problems are the handling of large volumes of solid rock, retorting, waste disposal, none of which has a counterpart in the crude business. "Oil executives just don't feel competent making decisions to invest hundreds of millions of dollars in a mining operation," says Russell J. Cameron of Cameron Engineers. "Oil companies are used to something they understand better," an official with Gulf Minerals Company, a subsidiary of Gulf Oil, explains, "and they'd rather bet their money on conventional exploration. It takes a good deal of evaluating and convincing to put money in a new direction and adjust to a new way." The author of a Shell report on shale noted, "The oil companies which could raise sufficient capital to start an oil shale industry prefer to risk their funds in more familiar activities, and the entire concept of mining is strange to oil companies."

(Ironically, similar reasoning appears to have deterred many companies with mining experience from becoming involved with oil shale. For example, it might appear that a company such as Koppers Company, Inc. of Pittsburgh would be a logical party to build a commercial shale oil plant. A large corporation with 1969 sales of $533 million, Koppers consumes sizable quantities of crude oil in its petrochemical operations and participates with Sinclair Oil, which recently merged with Atlantic Richfield, in a joint petrochemical venture called Sinclair-Koppers, Inc. Koppers also does a great deal of coal mining which brings in about a third of its revenues, and it possesses an experienced engineering and construction division which designs and builds plants for the steel and chemical industry. The firm has even had experience with oil shale as a consultant to the Bureau of Mines for its Anvil Points research facility near Rifle which the government operated during the 1940s and 1950s, and as designer of a proposed oil shale plant for the U.S. Navy. But according to Warren Riley, a Koppers executive who worked with the Bureau of Mines on oil shale, "We're just not an oil company. We're not in the oil business because it's just not our piece of cake. It's a business of its own, and we're not familiar with it, especially the facets of distribution and marketing." Koppers might participate with the oil industry in a shale project, he says, and in 1967 the company agreed to work with several oil companies and a couple of research organizations in a program which eventually did not get off the ground. "We do have something to offer the oil industry," Riley goes on. "We have competence in mining, solids materials handling, and we've built some large solids plants. These are things we're familiar with.")

There appears, too, to be an ingrained skepticism among many oil officials that anything could ever really be better than pumping oil out of the ground. "There is a gap between the cost of synthetic fuel and the cost of conventional crude and no amount of research is going to close the gap," argues Harold Weinberg, engineering head for the Synthetic Fuels Research Department of Esso Research and Engineering, a division of Standard Oil of New Jersey. "Look, when you talk about bringing the cost of producing shale oil down, you're asking for research to do in a relatively few years what the entire oil industry has done in a hundred years. And you want us to start with *rocks* instead of oil. It just can't be done. I don't care if we spend $100 million in 10 years or $200 million in 20 years. It will never equal the amount of effort that's gone into present petroleum technology, because you're going to always have to start with *rocks,* and you will never be able to start with *rocks* as cheaply as you can start with oil. Look, we're really interested in the North Slope. There seems to be a lot of oil up there, and although we don't know how much there is and how much money it will cost to get it out, that is *oil* that's up there, it's *pe-tro-le-um,* and the problems are ones we've faced before and that we've solved." Many oil shalers take a somewhat cynical view of this attitude. "If you gave an oil company a lumber forest, they wouldn't know what to do with it," says Dr. Tell Ertl, a former Union Oil official who now owns interests in several tracts of shale land. "They'd probably rig up a great big drill and start drilling the trees."

"Let's face it: producing shale oil is a dull, nitty-gritty business of digging up rock," Dr. Charles Prien of the Denver Research Institute says. "Exploring for oil is much more fun. It's adventuresome. There's nothing in oil shale as glamorous as sending the *Manhattan* through the Northwest Passage." Toward the end of a discussion I had with an executive from one of the major oil companies on the problems of producing shale oil, he suggested that "when you finish your book on oil shale, you ought to write about something that's really interesting: exploration for oil." His eyes gleaming, he commenced upon an impassioned depiction of the "romance" of the oil business, the excitement of drilling in the far corners of the world, of thrusting pipelines across unexplored mountains, of constructing offshore platforms to withstand hurricanes —all of which he felt was vastly more exhilarating than messing around with some black rock in Colorado.

In considering new ventures, there is a certain sheep-like tendency among the major oil companies, a feeling that there is no sense putting a lot of money into something like shale unless one's competitors are. "Why take all the risks of being first so all the other companies can take advantage of my mistakes?" a Mobil Oil executive asks. While oil companies like to keep up with what everyone else is doing, they tend to take the view that if they should really get behind in technology they can always "buy in" later if necessary. Walter S. Svenson, an executive with Standard Oil of California, testified at the BLM hearings that "We concluded that a good part of the information we needed would probably be available through the purchase of patents, know-how, and it wasn't in our best interest to embark on a major [oil shale] research program of our own." In 1961 Gulf Oil turned down an opportunity to purchase some shale land since, according to a memorandum in the

BLM evidence, "there will always be the possibility of buying in here or going elsewhere if and when shale becomes economic."

Of considerable importance to oil men are the financial differences between producing shale oil and crude oil. In a 1963 paper published in the *Western Economic Journal,* Dr. Henry Steele of the University of Houston estimated that the total cost per barrel of discovering, developing, and producing crude oil was, as of 1960, $4.15, which rises to $4.85 if a standard 14.5 percent royalty is added. The comparable shale oil per-barrel cost in 1962, he estimated, was only $1.46, and he said two oil companies told him their own estimates were lower. How could an oil company not afford to take advantage of this differential? Why were they not investing in shale rather than crude reserves? And for that matter why were they not losing money when the selling price of crude was around $3? The answer is that producing shale oil is a very conventional business: you build a shale oil plant and then try to sell a barrel of shale oil for a price far enough above your cost to allow you to earn an acceptable profit and return on your investment. The principal capital outlay goes into construction of the plant, which remains as a large fixed asset. Producing crude oil is very unconventional. The chief expense is just finding it, and large amounts, $800 million in the United States in 1968, are spent drilling dry holes. Steele estimated that the finding cost of a barrel of crude in 1960 was $2.00. But once oil is found, it can be brought to the wellhead quite cheaply. Thus while it might cost $2 million to discover an oil field containing $1 million worth of oil, the decision on whether to produce oil from the field has nothing to do with the $2 million exploration cost. That money is irretrievably spent and nothing but geological knowledge remains. The decision to produce depends only on the "marginal" or "incremental" operating cost of getting the oil out of the ground. If the marginal costs are sufficiently below what the oil can be sold for, oil will be produced even though, if the exploration costs were added, the field being produced would operate at a loss. This is because producing the oil would help to cut the loss.

Steele estimated that as of 1960, the marginal cost of producing a barrel of crude was only $1.18 against an estimated production cost of a barrel of shale oil of $1.46. Further, even the total cost of barrels being produced was much lower than his $4.85 figure, for the crude now being pumped comes from fields discovered an average of 12 years ago—the East Texas field is 40 years old and still being produced—when exploration and development costs were much lower. In the long run, Steele points out, crude finding costs have been rising sharply and will continue to rise as the nation's fixed stock of discoverable reserves decreases and new reserves become more expensive to locate. Anticipated shale costs, despite labor and material charges, are probably falling and should continue to fall as new technological developments occur. Yet unless an oil company were facing a serious crude shortage (most large oil companies are set up to refine and market their own crude production), it might have a strong motivation, Steele suggested, to opt for the immediately less expensive marginal cost of producing its existing reserves rather than investing in shale. And the more reserves a company possessed, the stronger its motivation would be.

Crude has other short-term advantages. L. L. Starlight, senior economist for Continental Oil pointed out at the 1965 Oil Shale Symposium that the low marginal production costs give oil companies protection against sudden drops in crude prices. Though crude prices have remained extremely stable over the past decade, he said a relaxation on controls over cheap imported oil could swiftly depress prices. In such an eventuality, a crude producer could merely cut his exploration budget for a time and pump profitably from existing reserves. A shale oil producer, who is always running on a full cost basis, might well be forced to operate at a loss.

The crude producer, finally, enjoys several widely publicized tax advantages, principally the percentage depletion allowance which permits him to deduct 22 percent (recently reduced from 27½ percent) of his gross income from his income tax up to a maximum of 50 percent of his net income. Shale is covered by depletion but the percentage is only 15.* Oil shale interests discuss this point continually. The Oil Shale Corporation calls it "gross discrimination against one group of producers as contrasted with competing producers of the same mineral product from other sources," though they do not mention that crude's depletion allowance is justified as a necessary means of stimulating exploration for new crude reserve discoveries, which is irrelevant to shale.

Another tax advantage allows oil producers to write off immediately certain "intangible" drilling costs, which substantially lessens the burden of exploration and development expenditures. An aspiring shale oil

---

* After many years of trying, Colorado congressmen managed to add a provision to the 1970 tax bill changing the point of application of the allowance from the mined rock, where its impact would be negligible, to retorted oil. Shale oil produced *in situ* might constitute an "oil well" under the law, some lawyers feel, and thus qualify for the full 22 percent.

producer, like most other businessmen, must invest after-tax dollars in his development work, and then amortize his investment gradually over a period of many years.

All of these factors mean that even if a shale oil operation could be proven to have an extremely high degree of profitability, so high that under normal circumstances a normal company would immediately pursue it, an oil company would have a strong motivation to demur. In the long run, of course, many of the cost advantages for shale become greater and greater. Still, oil companies may desire to take the attitude that they will think about such things as shale oil's potentiality for avoiding the cost squeeze from high exploration and development expenditures only when they really have to. "Had substantial oil shale deposits of the highest quality been held by firms outside the crude petroleum business," Henry Steele wrote in the *Western Economic Journal,* "development and production of shale oil would perhaps have occurred during the last decade."

This is not ignored by oil men. They are all too aware that while about 95 percent of the known oil and gas reserves, excluding Alaska, are located on private land, 85 percent of the shale is on public land. The shale in private hands, generally along the rims of the Piceance Creek Basin, is closer to the surface and more accessible. However the government shale is not only many times more extensive but it is much thicker, richer, and perhaps ultimately less expensive to develop—*in situ* techniques as well as huge open-pit mines would be especially amenable to the center portions of the basin. As Chapter Five will describe, the government has never formulated a broad-scale development program or announced any firm decisions on what it intends to do with its shale land. Under the terms of the Mineral Leasing Act of 1920, which prohibited further filing of mining claims on oil shale land, the Secretary of the Interior was authorized to lease public land containing oil shale, but he was left with considerable discretion in devising the terms and nature of the lease. Nobody knows, for instance, whether the lease would be designed to give the shale oil producer some of the tax advantages enjoyed by the crude producer. And, of course, nobody knows for sure who would end up getting the leases.

During the Senate Interior Committee's 1967 oil shale hearings, Colorado Senator Gordon Allott was discussing with Charles F. Jones of Humble the problems a company should consider before deciding whether to spend a lot of money on a commercial shale oil plant, and the following exchange occurred:

SENATOR ALLOTT: . . . A person or those companies with holdings in the southern part of the sector, along the outcroppings, would almost need to know for certain whether or not they could participate in the leasing and development of the deeper, thicker, and richer beds, would they not?

MR. JONES: Certainly I agree with that. I think, to make the principle here, Senator, what would be important to me is that I would like to have a little assurance about what the running rules of the whole game are going to be a little bit further down the road.

The worst possible position that a company can find itself in is to be trying to make investments for the future knowing that there are a number of factors that will be extremely important to it, and not having any notion as to what the running rules on these other factors are going to be.

Now, certainly in the case we are looking at here, with the vast public lands to the north of the lands you are describing, what is going to happen to these public lands is or should be of deep concern to the holders of private lands in the south.

So, although a person might not have to know exactly whether or not he could lease some of these Government lands, he would surely want to know what was going to happen to the public lands and what impact this would have on him from a competitive standpoint as the future unfolds.

The important point is that you need to know what the running rules are to have them nailed down, so you are sure of what environment you are going to have to operate in.

Colorado Senator Peter H. Dominick, who is not completely disinterested regarding oil shale development, told the committee that ". . . (O)ne could not reasonably expect private sectors to build a full-scale industry . . . on marginal lands lying on the fringe of the rich public deposits with the risk of being destroyed overnight should the federal lands be opened." H. Byron Mock, a Salt Lake City attorney who represents various mining interests and who was a member of the Oil Shale Advisory Board, stated in a 1966 *Denver Law Journal* article that during the board's discussions with oil industry executives the "fear" was often expressed "that private capital having been spent in the development and showing the way might give latecomers a chance to pick up federal leases and compete without the vast initial investments that appear to be necessary."

A number of oil men harbor a very special concern that the eventual developer of the public shale land could be the government itself. Indeed a number of people, such as former Illinois Senator Paul H. Doug-

las and University of Colorado Economics Professor Morris E. Garnsey, have suggested that a federal oil shale corporation modeled after
Comsat or TVA be formed. During the 1967 Senate Antitrust Subcommittee hearings on oil shale, Dr. Orlo Childs, president of the Colorado
School of Mines, former executive with Sinclair Oil and Phillips Petroleum, and a member of the Oil Shale Advisory Board, quoted the president of Humble Oil as telling the board:

Gentlemen, our company has grown over the years, and has become what it
is because we are willing to face competition, and we are willing to work
against competition, and welcome competition. At the present time we own
fee land in the Piceance Basin. But if the government were to go into business on its land in the Piceance Basin, the first thing we would do would be
to sell off our own lands because we will compete against anyone but the
Government.

"This is a fear, gentlemen," Childs said to the committee. "It is a fear
that many of these companies have. And I think you can see it for
yourself." H. Byron Mock concurred with this impression in his *Denver
Law Journal* article: "The fear that a government-operated oil shale industry might come into being after private industry had gotten started in
the less rich lands was a ghost that kept appearing."

Oil men are equally concerned over government research on oil
shale, and the following chapter will describe how the industry helped
close down the government's only major research facility. Charles F.
Jones of Humble told the Senate Interior Committee in 1967:

Humble believes that in the best interests of the nation, the mineral resources contained within the public domain should continue to be developed
by private enterprise under a minimum of Federal controls, as in the case of
oil and gas, and without further Government expenditures on research . . .
I cannot support the concept of government research in competition with
private research to satisfy consumer needs that are appropriately a part of
the private sector of the economy.

Later in talking about a possible federal leasing program, he expressed
concern "as to the role of the Government in carrying out research and
making information commonly available to all participants or all potential participants" in some future leasing program.

In *The Politics of Oil,* Dr. Robert Engler quoted Texas Governor Allen
Shivers as saying in 1954:

The problem in recent years has not been one of finding a substitute for a
limited natural resource but rather keeping ambitious bureaus of the Fed-

eral Government from substituting synthetic replacements before they are needed. Economic history shows that long before the depletion of any resource a dynamic and free people have replaced it with a better substitute.

Engler states that oil men fear a government-encouraged synthetic fuels industry might "provide a yardstick for judging and controlling costs and profits." Crude production costs are one of the industry's most closely guarded secrets, and the existence of a government-sponsored shale oil company would, at the very least, enable government in its leasing programs to be sure private companies did not achieve too high a profit.

All of this deliberation about who might eventually develop the shale needs to be put into proper perspective, however. We are not just talking about the future of the shale oil industry but of the oil industry in general. "Oil from shale, of course," noted the synopsis prepared by Interior for the Oil Shale Advisory Board, "would directly supplement or replace crude oil in nearly all its uses." "Liquid or gaseous products from oil shale will enter the market gradually, competing in certain geographic areas with marginal products from other sources," said a proposed oil shale development program outline prepared in 1967 by the Bureau of Mines. "But eventually the products from oil shale will dominate some markets—particularly the oil market." Every barrel of shale oil thus has the potential for displacing in the marketplace one barrel of crude oil.

To understand the significance of this potentiality, and its relevance to the oil industry's attitude toward oil shale, it is necessary first to examine in some detail how the oil industry operates.*

Fundamental to the economic health of the American petroleum industry is the domestic "posted" selling price of crude oil, which cur-

---

* Natural gas, while it supplies about a third of the country's energy requirements, is not an important factor in the competition between shale oil and crude oil and will not be considered in detail here. In contrast to the oil industry, the gas industry is considered by the government to be a public utility, and gas prices are tightly controlled by the Federal Power Commission. It should be noted, however, that gas is usually geologically associated with crude oil, and thus about 70 percent of domestic production is pumped by integrated oil companies. Any detrimental effects on crude production from shale oil might exacerbate a current serious domestic gas shortage, though the shortage is due not so much to an absolute lack of gas reserves as the preference of oil companies to spend their money on oil exploration instead of gas exploration. Due to low gas prices, oil companies figure their return on oil exploration is six times as high.

rently averages about $3.10 a barrel at the "wellhead." (The price varies from well to well depending on the location of the well and the physical characteristics of the oil). This price is not determined by the conventional capitalistic forces of supply and demand. It is a creation of a highly complex interplay of forces such as the oil industry's organization and structure, the operation of a maze of state and federal regulatory bodies and a system of tax subsidies, all of which act to keep the price artificially high. According to testimony before the Senate Antitrust Subcommittee in March, 1969, by Professor Walter J. Mead of the University of California at Santa Barbara, a typical barrel of Iranian heavy crude sells in the free world market for about $1.35. Adding 75¢ for transportation to a United States East Coast port and a 10.5¢ U.S. tariff, the total delivered price becomes $2.20. A comparable barrel selling at Refugio, a Texas coast port, for $3.12 costs 30¢ to ship to the East Coast, bringing its total delivered price to $3.42. Mead cited a January, 1969, Interior Department study which said that if the world crude market were allowed to exert a normal competitive pressure on the United States crude market, the price of oil east of the Rocky Mountains would soon fall about 95¢ a barrel.

This extra 95¢ is ultimately passed on to the consumer in the form of higher prices for gasoline, fuel oil, myriad petrochemical products such as plastics and clothing, and other products and services, such as ticket prices for airlines. Including the various special tax benefits received by the industry, the total cost to the public of the oil industry's preferential economic position made possible through state and federal law is estimated to run somewhere between $2 and $10 billion every year, depending on who is doing the estimating.* Dr. Mead before the Senate Antitrust Subcommittee looked at this cost in another way and called it "organized waste" or

. . . overinvestment in oil exploration and production, and consequent misallocation of the Nation's resources. We are developing resources at social costs of about $3.42 a barrel that have a social value of about $2.10 a barrel. Resource misallocation in turn results in a lower standard of living than is otherwise available to this Nation.

Oil executives admit these high prices entail some social costs, though they feel estimates by the industry's critics are generally exces-

---

* In its February, 1970, report to President Nixon, the Cabinet task force on oil imports put the cost to consumers in 1969 of import controls alone at $5 billion.

sive. But they maintain the extra cost is justified "to preserve a viable domestic oil-producing industry in the interests of national security," as the phrase often goes. The argument runs this way: Oil producers must receive sufficiently high prices for their crude, as well as special tax benefits from its production, so that they will have sufficient incentive to engage in the risky business of finding and developing new crude reserves. Otherwise, they might spend their time just producing the oil they already have. Our known reserves would then decline, and we would be forced to become more and more dependent on foreign oil. The nation would then become vulnerable to sudden cutbacks in its supply due to foreign political upheavals and perhaps even to various forms of political and economic blackmail. In wartime, we could be cut off from a material that now supplies (along with natural gas) 75 percent of our energy needs, and rapidly forced into surrendering.

It is not the purpose of this book to debate the merits of this argument, though most economists and others unconnected with the oil industry find it rather weak. It is relevant, though, to analyze just how the price of crude is maintained at its high level. Contrary to some allegations, it is not the result of conspiratorial plotting in hotel rooms. It is simply an ad hoc response by the industry, and, in turn, the government, to gradually developing business realities as well as the natural, understandable desire of any established business for self-enhancement. Once established, the institutional machinery began to operate automatically and it continues to operate that way today. "Nobody has to conspire," says a government lawyer. "That's the perniciousness of the beast."

One key to the system is the state "conservation" laws which evolved to prohibit waste of the nation's petroleum resources. During the oil industry's early years, the so-called "rule of capture" prevailed, which means that all the oil a man was able to bring to the surface from his well belonged to him. Oil resides in large underground reservoirs under pressure, and the first well drilled into a reservoir drains oil from the entire pool and releases the overall pressure, thus creating the familiar, but to oil men, odious geyser. Subsequent drillers not only found less oil under their wells but they had to install pumps to get it out since the pressure had already been dissipated. Oil drillers were therefore motivated to drill as many wells and to get as much oil out of the ground as quickly as possible. The result was oil production far in excess of market demands, which depressed prices, disrupted the market, drove many oil producers out of business, and, not incidentally, caused wasteful, inefficient utilization of the oil fields.

Awareness of these problems became abundant during the early years of the East Texas oil field when profligate production drove prices down as low as 10¢ a barrel. In a practice that was to become widespread in the major oil-producing states, Texas Governor Ross Sterling, former president of Humble Oil, instituted "prorationing," a complex production allocation and control system that came to be administered by the unlikely-named Texas Railroad Commission. There are two basic kinds of prorationing. "MER" prorationing is based on a determination of the Maximum Efficient Rate at which a particular reservoir, due to its physical characteristics, can be produced. Producers are then assigned "allowables" or maximum permitted production levels. "Market demand" prorationing is based on monthly "nominations" by the major crude buyers of the quantity of crude they intend to purchase that month. Generally, especially in Texas, the market demand is below the MER. In February, 1970, for instance, Texas producers were allowed to produce only 68 percent of their allowables, an all-time high record for the state, while Louisiana producers were limited to 48 percent.

This system was given special weight by the Connally "Hot Oil" Act of 1935 which prohibits interstate shipment of oil not produced in accordance with state conservation organizations. The prorationing states are also members of the Interstate Oil Compact, authorized by Congress, which coordinates their conservation operations. The purpose of the Compact is supposed to be to "conserve oil and gas by the prevention of physical waste thereof from any cause," and it specifically states that "It is not the purpose of this Compact to authorize the States joining herein to limit the production of oil or gas for the purpose of stabilizing or fixing the price thereof, or to create or perpetuate monopoly." Nevertheless, it is obvious that by matching supply to demand, the prorationing helps maintain price stability at a price level determined by the major crude buyers.

Clearly, the prorationing system can only be effective if the nation's overall crude supply is not overloaded with oil from sources not under its control. If there were too much supply from the outside, the conservation states would have to cut back production, with deleterious effects on local producers. Fortunately, the states which do not control production have a small output. California is an exception but it generally consumes more oil than it can produce. Offshore production in the Gulf of

Mexico is held down by Louisiana and Texas, even though the offshore lands are federally owned. Interior has been unsuccessfully trying to gain federal proration control of the offshore areas for several years.

The chief threat has been cheap imported oil. Imports first began to have an impact on the domestic market after discoveries of huge Middle East and Venezuelan fields around 1950. By 1952, as the foreign oil started trickling in, domestic producers demanded limitations. In 1954, the trickle became a torrent equal to about 10 percent of domestic demand. By 1957, the figure was 14 percent. Oil producers pressed their arguments of "national security" with vociferousness, and finally in March, 1959, after a rather unsuccessful plea for voluntary controls, President Eisenhower placed mandatory controls on imported oil and assigned quotas to refineries depending on their capacities and other factors—smaller refineries were deliberately given above-average quotas. The total share of the market to be occupied by the quotas was precisely established; the current figure is 12.2 percent east of the Rockies and the difference between demand and domestic supply on the crude-short West Coast. Because of imports from Mexico and Canada, which flow in under the auspices of other agreements and laws, and because of various other exceptions, the total market share of imported oil in 1969 was 22 percent.

(In February, 1970, a Cabinet task force headed by Labor Secretary George Schultz recommended to President Nixon that the quota system be abolished over a three-year period and replaced by a tariff system. Depending on which part of the world oil was being shipped in from, importers would have to pay the government a specified per-barrel tariff. Proceeds from the disparity between domestic and imported prices would thus accrue not to refiners but to the government, and might total $500 million yearly. The level at which the tariff was set would effectively determine domestic crude prices, and the task force recommended levels designed to force down domestic prices about 30¢ a barrel. The oil industry vehemently criticized the report, not just because of the recommended level of crude prices but because, said the *Wall Street Journal,* it "gives the Government too much control over prices. The industry fears the tariffs could be lowered any time in response to political pressure—a possibility which isn't inherent in the quota system." Some oil men also feared the new system would just be an intermediate step in the direction of eliminating all import controls. Walter J. Levy, who is considered by many as the dean of the world's oil economists, said its benefits could be illusory. Imposition of tariffs, he asserted, might well become "an immediate political trigger" for increases in taxes levied by major foreign oil-producing countries who would be reluctant to have the United States government obtain all the revenue benefits of the new system. As a result, he went on, world oil prices might rise as much as $1 a barrel, and oil companies would pass on their higher foreign taxes to United States consumers in the form of higher product prices. Shortly after he received the task force report, President Nixon announced he was postponing a decision on the recommendations indefinitely—probably until after the 1970 election in the view of some Washington observers.)

The overall pressure of the quota system on prices is not as great as it may appear. The Mexican and Canadian oil comes in at prices fairly close to domestic levels—the Canadian oil is about 40¢ a barrel cheaper—and the really inexpensive imports are spread around widely and flow to the smaller refineries, which are generally less efficient, so that the price differential is not truly reflected in significantly lower prices for refinery products. One government economist estimates that the true portion of the domestic market filled by cheap oil with a potential for depressing prices is probably around 5 percent. No one really knows how high this figure could go without affecting the posted price.

The susceptibility of a production limitation/stable price arrangement to disruption is demonstrated by what happened in the foreign market. Until the 1950s, it was under the control of the so-called "Seven Sisters"—Standard of New Jersey, Royal Dutch-Shell, Gulf, Texaco, Mobil, Standard of California, and British Petroleum. The Seven Sisters so effectively controlled production, set prices (which were equal to the posted price in Texas), and generally dominated the marketplace that a famous, highly critical 1952 Federal Trade Commission report dubbed them "The International Petroleum Cartel." The year 1952 was just about the height of the cartel's international influence, though, for soon widespread crude discoveries, a proliferation of independent producers, shippers, and refiners, the entrance of the Russians into the world crude market, and other factors all caused the Seven Sisters' market share to drop from 90 percent to 75 percent, resulting in a sharp drop in world prices. Now the cartel's power has been replaced in a different form by the Organization of Petroleum Exporting Countries (OPEC), composed of the ten largest oil-producing nations, which acts much like the Texas Railroad Commission to stabilize prices by controlling production. OPEC is not nearly as dominant as the cartel once was, but it does manage to keep world crude prices somewhat higher than they would be in a completely free market.

While the state conservation agencies are generally credited with being the major factor in stabilizing domestic crude prices at a high level, a number of government oil experts contend that as important, if not more so, is the structure of the domestic industry.

In theory, the oil business should be among the most competitive. There are a large number of producers, a large number of marketers, and the unit price of the commodity is low. Entry into the production end is simple and inexpensive. In fact, however, the domestic oil industry is dominated, even controlled, by about 20 large "integrated" oil companies which produce oil, transport it, refine it, and market it.

These integrated companies, generally termed the "majors," produce the bulk of their own crude requirements for their "downstream" operations, but about a third of the domestic demand is supplied by thousands of "independents." Independents sell almost all of their crude to integrated interests in the "field market" in the area of the oil field. The typical contract is an unusual device known as a "division order," which specifies only that oil be taken from a certain lease "ratably" with that from other leases in the area. Neither a fixed quantity nor a fixed price is mentioned; the price is ultimately determined solely by the big buyers. For example, crude prices rose around 5 percent in the spring of 1969 when some of the majors announced that as of a certain date they would start paying more for crude. Other companies soon followed suit, and within a few weeks after the first price hike (by Texaco) the *Oil & Gas Journal* reported that about 90 percent of the nation's crude was selling at the new higher prices, though the individual prices paid in specific areas by specific companies occasionally varied a few cents a barrel. According to a Justice Department Staff Memorandum prepared for the Attorney General's report in 1967 on "The Interstate Compact to Conserve Oil and Gas" (the Memorandum has achieved infamous notoriety among oil men almost equal that of the 1952 FTC report), "The field market for crude displays all the indicia of a monopoly market."

The majors' control is based in part on the fact that an independent's profits depend substantially on how quickly he can get his permitted oil out of the ground and sell it. His output has already been more or less matched against the desires of the buyers by market demand prorationing, and he is unlikely to want to pay high storage costs while he looks around for other buyers in the area who might be willing to pay him a higher price. The only alternative to an immediate division order would be for him to ship his oil somewhere by pipeline, by far the most economical means of crude transport. However, says the Justice report, "Crude oil gathering [shipment from the wellhead to the main pipeline terminal] and transport, performed as a single operation, is distinguished by the virtual nonexistence of nonintegrated interests." Though technically a "common carrier" regulated by the Interstate Commerce Commission, to which everyone should have completely free access, the pipeline system is owned and operated by the majors principally in the interests of the majors. Most represent "joint ventures" among the large companies, a friendly cooperative arrangement which extends to bidding on leases, whose prominence in the oil and gas industry has no parallel. An independent desiring to send an oil shipment through the pipeline system usually runs into a considerable amount of red tape, all of which takes time and costs money. Generally, rather than try to physically move their own oil through the pipelines, independents elect to sell their oil to the pipeline company and then purchase an equivalent amount at the destination point. This entails much less paperwork and fuss. "The importance of this in marketing terms," the Justice report says, "depends upon the power of the pipeline owner to insist that such a sale-resale transaction reflect the price basis which the owner has set for his own production and purchases." One of the reasons the international cartel broke down was that foreign transport is mostly by ship, which was completely unamenable to cartel control.

Beyond the field market, there is an active secondary market for crude, for many companies prefer to obtain someone else's crude which may be nearer to their refineries or more suited to their requirements, rather than bring in their own oil. The majors, who account for virtually all of the secondary market activity, do not buy and sell oil from each other, however. They *barter* and exchange equivalent amounts, with appropriate adjustments for differing qualities and transportation costs. The Justice report notes that bartering is less convenient than actual purchasing and selling, but is used in order "to leave the arbitrarily set posted prices the sole indicia of value."

Opportunity for disruption in the refinery area is minimal since, in 1966, 83.7 percent of the nation's refinery capacity was controlled by the top 20 majors, and concentration is rising, as it is throughout the petroleum business. As a practical matter, independent refineries would find it difficult to shop around the country for nonintegrated oil. To achieve maximum efficiency, a refinery requires a regular, continuous, guaranteed flow of crude, which can only be supplied through conventional channels.

Only after the petroleum products leave the refinery does the United States oil business become competitive. The marketing area contains a wide variety of independent businesses, though the price range at which they can compete is limited by the established refinery prices. In some parts of the country, the majors' domination remains high. Ten large oil companies account for nearly 90 percent of the gasoline market on the West Coast and 86 percent of the market in New England.

The *raison d'être* of this system, which is far more involved than can be described here, is that a high posted price of crude is quite important to an integrated company, more so than it might appear. In most businesses consisting of divisions that deal with one another, the price at which one division sells its output to another is immaterial. The chief consideration is how the company's aggregate costs compare with its aggregate revenues. To an integrated oil company, though, the official price at which its producing divisions sell to its other divisions is very material due to the nature of the industry's special tax credits. As has been mentioned, the percentage depletion allowance allows a crude producer to deduct from income taxes 22 percent of his gross income (figured on the basis of the posted price) up to a maximum of 50 percent of his net income. The initial rationale for the depletion idea when it was included in the 1913 income tax law—it was 5 percent in the beginning—was that a crude producer's oil well was a "wasting asset" and that he should be compensated for its steady loss in value as oil was produced and sold. Later expansion in depletion's scope and amount was justified by variations of the "national security" argument. The depletion allowance is frequently criticized, mainly on the grounds there is no limit to the amount that can be deducted from any one well's output (often it equals many times the original investment), that it applies to foreign production, and that by rewarding production from existing wells it actually deters rather than stimulates exploration for new wells. In 1968, some $280 million in depletion allowances was claimed by people, including many small investors, who did no drilling. The other main tax incentive applies also to production and is also justified on national security arguments. A crude producer may immediately write off "intangible drilling expenses" such as labor, materials, fuel, equipment repairs, which often total 75 percent of drilling costs.

While a dollar of profits from an integrated oil company's downstream operations is theoretically subject to the standard 48 percent corporate income tax, a dollar of profits from crude production, due to the tax breaks, may be virtually tax-free. Thus, according to the Justice report, it is in the interests of a fully integrated company to maintain a high crude price, even if its downstream operations were only to break even, so that it can make maximum profits at the producing end. The Justice report suggests this situation may have been a cause of the growth of integrated companies: anyone in the refining and pipeline business would want to obtain the tax breaks, as well as the assured source of supply, from producing his own crude; meanwhile anyone in the crude production business would want to annex downstream operations to reduce his vulnerability to a drop in crude prices, as well as permit him efficiently to dispose of his output. In practice, most integrated companies attempt to maintain a rough parity between production and consumption: in 1968, production of crude and natural gas liquids of 28 large oil companies equaled 96.9 percent of their refinery runs. Despite the parity, the importance of high crude prices is illustrated by an estimate in a recent *Forbes* survey that if import quotas were eliminated, and prices were to fall to the level of imported oil, the earnings of the average integrated oil company would be cut about 25 percent. To the extent that an integrated company is forced to buy crude from independents its profits are theoretically cut back, though sales are often arranged at informal discounts. But Dr. Alfred E. Kahn of Cornell told the Senate Antitrust Subcommittee in March, 1969, that if product prices were to go up only one-half of crude oil prices, an integrated company would still profit on balance as long as it produced more than 40 percent of its crude needs.

Exactly how much the oil industry really benefits from all these controls and tax advantages is a matter of much debate. Critics often emphasize the minuscule federal income taxes paid by oil companies. Between 1962 and 1967, Atlantic Richfield (before merging with Richfield) paid no income tax whatsoever on its $500 million in net profits. Between 1963 and 1967, Jersey Standard, Gulf, Mobil, and Standard of California paid 4.9 percent on their net profits of $21 billion. The overall industry average is higher: 21.1 percent against 43.3 percent for all manufacturing. Foreign oil-producing countries have taken advantage of this by levying very high royalties—the companies call them taxes so they can be deducted—on oil produced in their countries, which range as high as a dollar a barrel or more, and they continually are seeking to hike the rate if they feel the companies and the market can stand it. As a consequence, the profitability of foreign operations to United States oil companies has been steadily declining and, according to a Chase Manhattan Bank study it is now below that of domestic operations. One effect is that the United States has been stimulating foreign oil production at the expense of its own.

The state conservation system is also something of a double-edged sword, for it has caused gross overinvestment and manifold inefficiency and waste. For example, by giving each well an allowable it stimulates flagrant overdrilling. And it favors production of inefficient wells over efficient ones. According to a *Fortune* study in 1965, conservation practices cost the industry between $2 and $4 billion annually, which causes production costs to rise, which causes exploration and new drilling to decline. Dr. Alfred E. Kahn told the Third Management Conference on the Economics of Petroleum Distribution in 1967 that:

> The oil industry mistakenly regards as advantages the proration of crude oil and import controls. . . .
>
> Special tax privileges, like the depletion allowance, lead to heavy investment, which leads to excess capacity, which leads under the prorationing system to cutbacks.
>
> But these cutbacks then lead to increased operating costs, which lead to higher prices, which bring on excessive drilling for more moneymaking opportunities.
>
> This boosts costs, which compels domestic companies to seek lower operating expenses abroad, which results in increased oil imports, which in turn brings cries for government protection from domestic companies.

The result, Kahn told the Senate Antitrust Subcommittee, is that capital will continue to flow into the oil industry "until the cost burden of excess capacity is just sufficient to eliminate the artificial stimulus to investment that created it in the first place" and "profits are reduced by the low levels of capital utilization just enough so that new entrants no longer see the likelihood of earning supernormal profits." High prices only cause higher costs, he said, and costs are equated with prices rather than the other way around.

For these reasons and others, the oil industry's reported average return on net worth was 12.9 percent in 1968, very close to that of other manufacturing industries, though some critics contend the industry manages to hide at least some of its profits. Measured from the standpoint of cash flow, which includes net income plus depreciation and depletion, the oil industry is superior to others.

The effect of various companies' particular interest in the domestic price of crude is well illustrated by the continuing battle over possible revisions in the import control program. Avidly in favor of increased imports is the large petrochemical industry which, except for those units which are parts of integrated oil companies, must buy its crude feedstocks on the open market. (The industry favors, however, tariff restrictions on some chemical imports.) A number of large chemical companies were granted special import quotas by the Johnson Administration when they threatened to build their new plants overseas to get access to cheaper foreign oil, and recently the industry has been renewing these threats. Also favoring greater imports are "crude-poor" integrated companies and independent refineries. Occidental Petroleum created a *cause célèbre* when it proposed to build a huge refinery near Machiasport, Maine, if it were given permission to create a free trade zone and bring in 300,000 barrels/day above and beyond the quota system. Occidental altruistically allowed as how it wanted to lower high New England fuel oil prices by 10 percent, but its chief motivation was to be able to bring in more of its extensive reserves in Libya—considering the constant threat of expropriation by host governments, companies with extensive foreign holdings like to get as much of it out of the ground as fast as possible. Occidental couldn't care less about the domestic market since it produces virtually no oil in this country. Small independent crude producers with no foreign reserves, whose livelihood depends most critically on high crude prices and the maintenance of the conservation system, are understandably violently against Machiasport and the lifting of import controls in general. The large international companies with large reserves both at home and abroad are somewhat in the middle, but because of declining profitability abroad they are more anxious than they once were about the level of domestic crude prices.

Similar reasoning determines the attitude toward oil shale by oil companies, for the industry's stance is far from monolithic. Companies with large foreign reserves are generally least interested in shale development, and should a domestic crude shortage develop, they would be much more inclined to have import controls lifted than develop shale. Virtually the only oil companies, with the exception of Atlantic Richfield, to show a genuine interest in shale development are crude poor. Not only do crude-poor companies have a continuing concern that a crude shortage could lift prices and catch them in a cost squeeze or perhaps make it impossible for them to get all their crude requirements, but they have relatively little stake in maintenance of the present high crude price. Standard Oil of Ohio, whose interest in shale development has been genuine and ambitious, produces less than 25 percent of its refinery feed.

It should be clear from the foregoing extended discussion that in order for the price of crude to be maintained, it is necessary for a regulated supply of crude to be funneled through controlled channels before it can emerge into the relatively competitive consumer and business end-use markets. The system is threatened if a sizable amount of oil is produced outside it. It is threatened even further if this oil can somehow reach the product marketplace without flowing through established channels. This is precisely the threat posed by a thriving shale oil industry. As Robert Engler noted in *The Politics of Oil:* "The oil industry's basic fear is of a government-encouraged synthetic-fuel program that could challenge or upset the entire industrial government of oil with its careful balance of supply and demand."

Despite the current lack of activity, it is likely that a shale oil industry could be stimulated into existence by either construction of a large shale oil plant of undeniable profitability or an ambitious federal development program. If the major oil companies then rushed in and quickly assumed control, the future of the fledgling industry is difficult to predict. It would likely expand in accordance with the oil industry's actual need for shale oil to supplement its crude supplies. In any case it would be blended into the conventional oil system in such a way as to pose the minimum disruption.

But if the new shale oil industry were somehow taken over by independent business interests with no stake in the health of the crude oil industry, the following scenario, suggested in part by Dr. Henry Steele in his *Western Economic Journal* article and his testimony before the Senate Antitrust Subcommittee, is conceivable:

At least during its first few years, the incipient shale oil industry's production of perhaps a few hundred thousand barrels/day will be easily absorbed by ever growing crude demands without any effect on crude prices, and shale oil producers will of course be interested in getting the maximum price for their oil. If necessary, the response of the state conservation groups will be, said Steele, to " 'make room' for shale oil just as they have 'made room' in the past for imported oil." However, Steele added, "downward pressure on regional prices will eventually develop," and as the rate of shale oil production increases and causes the beginnings of an oversupply of oil on the market—perhaps exacerbated by substantial production from the North Slope *—the conservation organizations' willingness to "make room" will decrease, and pressure for reform of the conservation setup may grow. If reform occurs, a great many inefficiences will be eliminated, lowering the cost of crude production and making possible a modest drop in crude prices, which may make it difficult at least temporarily for shale oil to compete profitably. The result will be perhaps a slightly lower level of oil industry profitability, and probably the elimination of many small high-cost crude producers, but essential maintenance of the current system.

But if the posted price of crude is forced to remain at a lower level, there will be damage to the oil industry's assets. Dr. Walter J. Mead told the Senate Antitrust Subcommittee, in the following exchange with S. Jerry Cohen, the committee's chief counsel:

DR. MEAD: . . . (T)he company that has vast reserves of oil in the ground, crude rich, has a resource, the value of which depends on the price of oil. The value of any resource in the ground, call it economic rent, is a function of the value of that product in markets minus the cost of getting it there.

If a vast shale industry is opened and we get a lot of production, new production, from oil shale, this resource is so vast and so staggering to the imagination that it is capable of depressing the price of crude.

Now if the price of crude should, therefore, go from, say $3 a barrel down to $2.50—that is barrels in the market, not in the ground—it would correspondingly reduce the value of oil in the ground by 50 cents, and that is a loss in economic rent. This is an important issue to crude-rich companies. . . .

* * * * *

MR. COHEN: It has also been suggested that crude-rich companies also may have the incentive not to have others develop this land for fear of what it might do to crude prices in this country.
DR. MEAD: Yes.
MR. COHEN: Do you have any comment on that?

* A November, 1969, *Wall Street Journal* interview with William Fraser of British Petroleum Co., which owns an estimated 4.8 billion barrels of Alaskan oil, said he felt that market demand prorationing on the North Slope was a "possibility."

DR. MEAD: Well, the preliminary statements or studies that are available on cost—cost of producing oil from shale and revenue indicating that the industry right now is very profitable, and not just a little bit—it is very profitable. You had Professor Henry Steele testify here. He has what I believe to be one of the best studies available. And his studies do indicate that it is very profitable.

Now, if it is, and if this industry consequently gets started in a big way, we are talking then about a large increase in the supply of oil in the American market. And given that, any further price increase for crude seems to me to be out of the question. Prices are likely to go down. And as I indicated earlier, any firm with large reserves is not terribly interested in seeing the price go down.

MR. COHEN: Which means not having the resources developed by anyone.

DR. MEAD: That is correct.

Steele suggested that a not too uncomfortable stabilization of the oil industry at a lower crude price level may not be possible, especially if the state conservation groups are unable to reform themselves. "If [crude production] costs cannot be cut," he said, "and if shale oil costs do not increase, allowables will have to be cut more drastically, economic and political repercussions will increase, and the state conservation system might conceivably break down entirely."

Eventually, as the shale oil industry continues to expand, assuming no serious technological hitches, production will attain sizable levels, perhaps several million barrels a day. Since existing pipelines in the area have a capacity of only slightly over a hundred thousand barrels/day, shale oil producers will probably build their own pipelines to the West Coast and Chicago markets, and they may even build their own refineries specially suited to handle shale oil. Meanwhile economies of size will continue to bring down costs and the pressure on crude prices will continue, though, ironically, as long as crude prices remain high, shale oil producers will be benefiting from the very system organized for the crude industry's benefit while at the same time using it against the crude industry. But if crude prices continue a general decline, crude producers will be forced to compete by cutting back on exploration and producing from existing reserves at a low marginal cost. The North Slope may be a help to crude producers at this time if the deposit proves to be really large and if production and transportation costs turn out to be low. The ability of the shale oil industry to continue its pressure will depend, again, on the level of profitability it is willing to accept and its ability to cut operation costs. If it continues to thrive, the oil industry will find itself in the ignominious position of running out of crude reserves and being unable to finance the discovery of new ones.

Before this can happen, the major integrated companies with foreign holdings may well retaliate by pressing for unlimited foreign imports, for potentially the big low cost foreign oil fields could be produced at a high rate for many decades. The competitive situation will then depend on the level of imports, with its accompanying unfavorable balance of payments implications, the nation will accept, the level of royalties the OPEC will demand, the level of the free world crude market price—which oil companies claim would rise if import controls were relaxed —and, of course, the level of shale oil production costs. Even if imports dominated the East Coast, it is likely that shale oil would continue to prevail, because of high transportation costs, in the mid-continent and West Coast areas.

If increased imports are politically unfeasible, it is possible the oil industry will demand some kind of protective legislation to put controls on shale oil production. Governmental protection of an indigenous oil industry from synthetic oil is already being practiced in Alberta, Canada. Underneath bleak, muskeg-covered land 220 miles north of Edmonton are an estimated 300 billion barrels of oil contained in what is called the Athabasca tar sands. These black sands contain bitumen, a thick asphaltic material believed to have been deposited by conventionally-formed oil reservoirs that for some reason flowed out over several thousand square miles of sand and clay. Some of the tar sands lie near the surface, but like the oil shale much of the deposit lies under hundreds of feet of overburden. A number of oil companies have been interested in developing the tar sands, notably Sun Oil, an extremely crude-short company which in 1968 produced only 46 percent of its refinery runs. The Alberta government, which owns most of the land, was for many years reluctant to permit development for fear it would upset the market for conventional crude produced from the oil fields in the southern part of the province. Alberta has excess production capacity and imposes strict production controls. Finally Alberta's Oil and Gas Conservation Board permitted Great Canadian Oil Sands Ltd. (GCOS) in which Sun holds an 81.3 percent interest, to begin production, but established a production limit of 45,000 barrels/day, about 5 percent of the province's conventional oil output. The government explained that its principal reason for establishing this policy "was to provide for the orderly development of the oil sands in such a manner as to supplement but not displace production from the conventional industry."

Sun began producing in October, 1967, using the "hot water" method of recovery, in which hot water is bubbled through strip-mined tar sands, carrying organic materials to the surface while precipitating the sand. The process is more familiar to oil men than shale oil production since it is related to techniques used for secondary recovery of oil from conventional oil fields—geologically there is very little difference between viscous oil often encountered in conventional fields and tar sands. Sun has run into serious cost problems, however, and original estimates of $160 million have soared to a reported $325 million.

Meanwhile, an organization called Syncrude Canada Ltd., jointly owned by Atlantic Richfield (30 percent), and Canadian subsidiaries of Cities Service (30 percent), Standard Oil of New Jersey (30 percent), and Gulf (10 percent), had been repeatedly petitioning the Alberta government to let them build a plant, too. The plan was hotly opposed by many other Canadian oil companies with greater interest in the province's current production, and their concern was not mollified by the North Slope discovery which could have a large effect on the level of imports Canada will be allowed to send into the United States. After Syncrude agreed to a four-year postponement, to 1977, for beginning operations, Alberta granted it in the fall of 1969 permission to build a $200 million plant which will produce a maximum of 80,000 barrels/day of petroleum products including 50,000 barrels/day of synthetic crude. The combined 125,000 barrel/day output of GCOS and Syncrude will still be a relatively small factor in the Canadian market, whose demand in 1977 including exports is expected to be at least 3.5 million barrels/day. Several other groups have considered applications for tar sands plants—notably the government-owned Japan Petroleum Development Corporation. However, the Alberta government can be expected to make sure that oil from tar sands will flow into the market with minimum disturbance of the established crude industry.

Apparent sympathy for the Alberta government's viewpoint was expressed by former Interior Department solicitor Frank J. Barry at the 1965 Oil Shale Symposium:

The specter of [a] community of poverty-striken Texas oilmen resulting from a dislocated oil industry [caused by widespread shale oil production] may have its amusing aspects to some, but should it occur, the entire Nation, as well as the industry, would suffer the burden of relocating a huge sector of our present economy.

Henry Steele, however, expressed doubt that the oil industry's demands for shale oil production controls would be met: "The argument that the domestic crude oil industry should be protected from foreign competition in the interests of national security can scarcely be advanced against the competition of domestic shale oil." No case can be made for the application of controls from a conservation standpoint, he pointed out, since the shale reserves are all but limitless. "Consumer interests," he said, "should be alert to possible attempts to limit shale oil output in the interest of 'conservation.' " But even if a strong case could be made for easing the wounds of the oil industry, Steele added, "Such a policy is not without certain costs if its implementation means suppressing the rate of introduction of a lower cost energy source in favor of maintaining markets for higher cost sources."

It is possible to debate and speculate at length about just how much damage the oil industry would suffer in an all-out battle with shale oil. Even slight damage may be very unlikely in view of the several fallback levels available to the oil industry, each of which might subject the shale oil producers to a serious profit squeeze. The industry's sense of self-preservation, further, is not inconsiderable, and as a government lawyer notes, "We've tried hard, but we've just never been successful in breaking up their arrangements. They can create antitrust problems faster than we can solve the ones they created 50 years ago. Your suggestion on oil shale is probably just a lovely dream."

But trouble from shale oil is unquestionably a possibility, and oil men can be sympathized with if they view it with some apprehension. More than executives of any other industry, they are used to the comfortable maintenance of a tightly controlled, government-assisted status quo, with only the most vestigial trappings of competition. This attitude was well demonstrated in an exchange in May, 1969, during Senate Antitrust Subcommittee hearings on federal oil import policies between Senator Philip Hart and Kenneth E. Hill, senior partner of Eastman Dillon, Union Securities & Company, who runs that firm's investment banking relationships with many oil companies:

HILL: If by happenstance import quotas were removed and the tax incentives were removed, there is no doubt at all that most of these oil companies would show substantially lower earnings. Therefore, their stocks would sell lower and the asset values of mutual funds and pension funds would be reduced.

Now, this is just a side effect. I just wanted to bring before the committee the fact that it is an enormous industry, that investors and management have made huge capital expenditures year after year on the presumption that these—the framework we now work in would continue for some time.

I am on some boards and I see these long cash flow schedules and they all just assume the present price, present intangibles, present inport quota, and they all say we will get a 12-percent return on our investment, so let us make the investment, and shareholders look at the balance sheet or the annual statement of a large company and see they are earning 12 percent and growing a little bit, so they say, I would like to buy 100 shares.

All these things would change, particularly if you do not know the rules of the game.

I think any changes that might ever be contemplated should be quite gradual so that all concerned—everybody in the United States is mixed up in this huge industry. Everybody's welfare is involved, both from national security and the national economic welfare.

I think any changes, therefore, just have to be slow and gentle over many years.

SENATOR HART: Well, it is, you will agree, an interesting philosophical problem.

MR. HILL: Yes; it is. I do not think the Government should subsidize or guarantee any stock investor any kind of return. That is up to the management.

SENATOR HART: I thought that is what you just said.

MR. HILL: What did you say, sir?

SENATOR HART: I got the impression that that is exactly what you thought should be done.

MR. HILL: No. I did not say guarantee. I never said any guarantee.

SENATOR HART: Subsidize?

MR. HILL: That is your word.

SENATOR HART: I thought you said subsidize or guarantee.

MR. HILL: I just said maintain the present economic framework within which the industry operates. You can call it—I mean other people call it subsidy.

SENATOR HART: I do not want to put words in your mouth if you have not said it. I thought you said it.

MR. HILL: I hope I did not.

As long as the "economic framework" remains carefully fixed and controlled, oil men of course are willing to compete quite briskly with each other. They are willing to bid energetically for leases. They are willing to joust with such uncertainties as how much oil there is on the North Slope and how much it will cost to get the oil out. Some of them are even willing to consider producing tar sands since the rules and policies are clearly stated and the Alberta government is committed to protecting the existing system. But in contemplating oil shale, a vastly alien substance to begin with, they see nothing but an incredible array of unknowns. Many are relatively simple: What will be the depletion allowance on shale oil? What kinds of import quotas will be granted to shale oil refiners? What will be the royalties and other provisions if the government grants leases? Other unknowns are more complex: Will the government conduct a big research and development program on oil shale and make the technology widely available? Just how easy will it be for non-oil companies to get into the shale oil business? Who will the shale oil producers be?

But the biggest unknown of all is: *Will the oil industry be in control?* The oil industry may now possess great political and economic power, but power in American society can be surprisingly ephemeral. Seemingly impregnable one moment, it can become strangely vulnerable the next. Few oil industry critics believed at the beginning of 1969 that only 12 months later, they would succeed in knocking 5½ percentage points from the industry's mighty bulwark, the percentage depletion allowance. It is understandable that oil men might want to retain the status quo on oil shale or perhaps even, as Henry Steele said in his *Western Economic Journal* article, "retard the development of oil shale." Suppose an oil company actually went out and bravely constructed a commercial shale plant? "It is well within the realm of the possible," commented a 1968 report by National Economic Research Associates, Inc., a Washington consulting firm, "for the petroleum industry to find itself unintentionally the godfather of a fledgling shale oil industry which, once started, could begin the function of supplanting its predecessor." Robert Engler wrote in *The Politics of Oil:* "Oilmen deplore the conducting of industrial [synthetic fuel] experiments that might easily lead to full-scale operations, just as they are wary of any federal research or support that ultimately could threaten their patent-based controls over petroleum."

Henry Steele suggested in a 1968 article in *Natural Resources Journal* that the probable heavy concentration of a future shale oil industry plus its location in no more than three states could lead, if the production of oil from shale were substantial, to "a serious weakening in the political power of the industry":

Political influence does not depend nearly as much upon the financial power of the few major firms, as it does upon the large number of small producers, the great number of royalty recipients, and the widespread geographical incidence of oil production among the states. Moreover, while crude oil is currently produced in over thirty states, shale oil will be produced in at most three states, and possibly in only one for quite a while. A state's position on oil industry legislation is curiously affected by the circumstances of its having some crude oil production, even though it may consume vastly more than it produces. If for no other reason this is because producer lobbies are much more active and effective than consumer lobbies. In the crude oil industry, producing states are not only numerous, but include almost all of the politically most influential states; in the shale oil industry, producing states are not only few but without much political influence at the national level. . . .

. . . [T]here will be no lobby of several million royalty recipients with the political power to influence the legislative fortunes of the shale oil industry. Shale oil production will therefore be more vulnerable to shifts in administrative and legislative climates than the crude oil industry has been.

"The oil industry doesn't really want any action on oil shale," contends Stewart Udall, who spent eight years as Secretary of the Interior and now is head of The Overview Group, a Washington-based environmental consulting organization. In attempting to devise an oil shale development program, Udall says he found in his discussions with oil industry executives that the potential competitive threat from shale oil "looms very large in their minds. I never sensed any real driving interest in shale development. If the oil industry really wants to go somewhere, they have plenty of money to spend. But all they wanted is some shale land as a hedge in case something happened. Look, shale development might alter the foundations of the whole oil industry. They think it's fine as long as everyone plays according to the rules. But when something comes along that might change the rules, then they are very wary of it." Udall's own views on who the new rulemaker might be were perhaps indicated by his statement to the Senate Antitrust Subcommittee in 1967 that oil shale "will enable us, if we do it right, in my judgment, to have a national energy policy that is much sounder than the one we have today, for it will enable the Federal Government, because the people own this resource, to guide the energy policy of the country down the right track." He did add a little later, however, in talking about the Alberta government's tar sands policy, that "they can control it, just as we would be able to control oil shale—so as to produce the minimum disruptive effect on the energy economy."

It is problematical just how many oil men would be willing to place their trust in the federal government in this regard. To them, compared to the comfortable stability of the petroleum cartel, oil shale is a game with extremely uncertain, unpredictable rules. The way oil men view participation in such a game is well illustrated by this exchange during the Bureau of Land Management hearings between Interior lawyer Robert Mesch and R. G. Christianson, Shell Oil's executive vice-president for exploration and production, who had characterized Shell's current shale research efforts as "very small" and "quite modest with respect to the company's overall research":

CHRISTIANSON: . . . I think in answering your question I don't quite know what rate of return on an oil shale operation might be.

MESCH: Why is this?

CHRISTIANSON: Because of the many unknowns that at least at our point of time are involved in the—one can make speculations if this happened, this might be, and if under this condition, it might be that, and one can establish some sort of range that might be from nonprofitable to extremely attractive, but one couldn't say what it will be at this point of time.

MESCH: What are some of the unknowns?

CHRISTIANSON: Simply the entire cost, the entire royalty, the depletion treatment, the handling of materials, the quality of materials that one gets out of some sort of an operation, finally, how it enters your stream, what the refining costs are, these are not absolute figures at this point. This is the type of thing that one makes further investments in in order to generate and establish costs. This is the sort of thing I speak of in terms of additional research.

* * * * *

MESCH: So far as you are concerned, you feel that Shell is in no position to really judge today whether the rate of return on an oil shale operation would or would not be comparable with its rate of return from other investment opportunities.

CHRISTIANSON: Quite frankly, there are some aspects of it that we just don't know the absolute numbers on, and it still boils down to the fact that we are not sure enough that we could make a prediction about what we might expect out of it.

MESCH: And could this prediction be made if additional research work was done such as in mining and retorting?

CHRISTIANSON: This is the essential sort of investment that would be being made. Again, with the land situation and those problems, which are outside our control, being resolved, such matters as how the depletion treatment would be, what indeed might be the price of a barrel of shale oil, these sort of factors, the economic environment in which this industry is going to live, if those matters beyond our control would be stabilized and we could indeed have the opportunity to address ourselves to those things that we might do within the industry and within our technology to either enhance our pursuit of this business or decide that it was something we wouldn't get into.

Chapter 3:
The oil industry's enthusiasm for the few oil shale research projects which have been undertaken has not been overwhelming.

The visitor to Rifle, Colorado, rapidly gets the feeling he is looking not at the birthplace of an oil shale industry which someday in the future will flourish mightily but at the skeletal remains of an industry which expired long ago. The land at the base of the oil shale cliffs and the bottoms of canyons leading toward the center of the Piceance basin is littered with the debris of unrealized dreams: rotting wooden tramways, rusting pipes, and crumbling stone oven-retorts. The old mines in the Mahogany Ledge have partially collapsed, and scattered about are shale fragments, bleached white by the sun and air.

A few miles southwest of Rifle, a narrow road juts off from Route 70, the main highway that follows the Colorado River from Rifle to Grand Junction. From the middle of a sleepy hamlet called Grand Valley, it winds eleven miles toward the cliffs through green farmland at the bottom of a canyon carved by Parachute Creek, eventually reaching a huddle of dirty, corrugated metal buildings behind a chain-link fence. On one of the buildings a fading orange sign announces: UNION OIL COMPANY OF CALIFORNIA. Standing on a plateau overlooking the buildings is a silvery structure of pipes and girders that shines brightly in the sun. The structure is only a frame, though, for most of the intricate inner machinery, which at one time was used to retort oil shale, has long since been torn down and sold for scrap. A few dozen black oil barrels lie in neat rows, and weeds poke through the shale-strewn ground. In a building, an elderly watchman sits at a dusty desk, his eye on an electric wall clock. A few sheep graze nearby.

"Essentially," says Walter Barnet, who heads Union Oil's oil shale research efforts, "the facility is on a standby basis."

The Union Oil Company of California has historically been the only major oil company with a continuing interest in oil shale. During the 1916–1920 boom, Union was the first to acquire shale land—it now owns 37,116 acres—and during the 1920s Union engineers made detailed studies of some 20 operating retorts. The company's interest subsided during the glut period of the 1930s, but in 1942 it once again began experimenting with shale oil extraction methods, primarily as a result of the urgings of Albert C. Rubel, a trained geologist who was Union's vice-president for exploration and production. By 1948, a 50-ton/day "semi-works" retort was operating in Wilmington, California, and in 1950 a public demonstration was made to numerous government and industry officials. Little more was done, however, until October, 1954, when Dr. Tell Ertl, who had worked as a mining engineer and geologist for the Bureau of Mines and for Union between 1948 and 1950 and was at the time chairman of the Department of Mining and Petroleum Engineering at Ohio State, urged Union to set up a shale oil committee to study the situation. The committee was formed—Ertl was a non-voting "consultant"—and after a study made its presentation to the board in January, 1955. It stated that on the basis of a 15,000–50,000 barrel/day plant costing $50–$150 million, "We believe that as a source of liquid fuel, oil shale is now competitive with domestic petroleum. Our calculations indicate earnings may be 10 percent after taxes and that technological advances could increase those earnings to as much as 20 percent." The report pointed out that "an average investment in finding, developing, and producing new crude oil in this country will earn today in the order of 10 to 12 percent after taxes." Union at the time was earning 8.6 percent on its net worth. In May, 1955, Union Vice-President W. L. Stewart, who was also a Shale Oil Committee member, stated: "Certainly if the rising trend of finding costs for petroleum continues upward, and there is no reason to believe it will be reversed for the long pull, and the demand for petroleum type fuels continues to increase, then it follows that products from petroleum must increase in cost, and this may give oil shale its great opportunity." Union also was a relatively crude-poor company—in 1954 it bought over 30 percent of its requirements from other producers—and it operates in the West Coast market where supplies have generally been tight and exceeded by demands. In many ways, the market can be considered as quite separate from the rest of the country, as its special consideration under the import quota laws indicates.

After the Shale Oil Committee's presentation, the board approved a $3 million appropriation for a "demonstration" plant of "semi-commercial" size. The aim of the project was summed up by the title of a speech made by W. L. Stewart: "A $5,000,000 Question: Can We Produce Merchantable Shale Oil at a Cost Low Enough to Make It Competitive with Petroleum?" Construction began at the Parachute Creek site. Union's retort was a giant structure four stories high designed to process 360 tons of shale per day. Raw shale was fed into the bottom by a "rock pump" while air preheated by hot shale ash was pulled down through the rising shale bed by a blower. Additional heat was supplied by the residual coke. A key feature of the idea was that the shale oil vapors flowed through the rising shale. They were thus condensed and cooled to near atmospheric temperatures while the shale was being preheated. No cooling water was required.

Dry runs began in February, 1957, and on May 18, 1957, with loud fanfare, considerable national publicity, and the presence of three governors, the plant was formally dedicated. In a story in June, *Time* magazine quoted Albert C. Rubel, who had become Union's president the previous year, as saying, "We have about $15 million in the pot, and that's only the ante." In *The Black Bonanza,* a company-approved history of Union written by Earl M. Welty and Frank J. Taylor, and published by McGraw-Hill in 1958, Rubel is quoted as declaring after the retort's first six months of operation that "It looks like we're on the right track. It has tremendous possibilities. We're already approaching the original assumptions of our research staff." The book also stated that "the retort exceeded the most optimistic predictions of its builders averaging in one forty-day run over 300 barrels of crude oil per day from 500 tons of shale."

By January, 1958, Union had begun to sell shale oil commercially to a number of mining companies for use in boilers and roasters. On June 30, 1958, a presentation about the "Shale Oil Project" was made to Union's board of directors which said that considerable progress had been made. Among other things, the rock pump idea had proved "entirely practicable." Further, "much progress has been made in simplifying the retort's design, and, consequently, reducing its costs. . . . Union's retorting process has been developed and proven to be a technical success." Finally, " 'Commercial shale oil' is a partially refined shale oil product that can be transported by pipeline, blended with crude oils, and sold as a high grade crude oil. It is superior in refining value to most of the world's crude oils. . . . Products from shale oil were indistinguishable from those produced from the best crude oils."

Nevertheless, at that meeting the board voted to close down the plant, on which a total of over $12 million had been spent, and a public announcement was made the following day. The company explained it wanted to analyze the data it had obtained. The plant has never resumed operations and in 1960 much of the retort machinery was torn down.

In public speeches over the past few years, Fred L. Hartley, formerly head of the Shale Oil Committee who took over from Rubel as president, has offered two principal reasons for the close-down. During 1958, he told the Senate Interior Committee in 1967, oil imports "were coming into this country with no regulations whatsoever, creating an oversupply of oil." The other reason, he said, was oil shale's low depletion allowance compared to crude. Hartley wrote a letter in 1963 which is contained in the evidence for the Bureau of Land Management Contests that states: "In my opinion there would already be a large scale shale oil plant in operation if the economic environment had been made equivalent to the tax treatment afforded conventional crude oil production."

Neither of these explanations seems wholly adequate. It is true that oil imports had risen from 10 percent of domestic demand in 1954 to 14 percent in 1958. However, as related in the previous chapter, President Eisenhower had established voluntary import controls in 1957 and in March, 1959, eight months after Union's plant was closed, had instituted mandatory controls. West Coast crude prices did drop during 1958, but only after they had risen artificially in 1957 due to the Suez crisis. According to Bureau of Mines figures, the average per barrel wellhead price of California crude was $2.62 in 1956, $3.05 in 1957, $2.90 in 1958, and $2.55 in 1959, though discounts from these posted prices were often available. Over the past ten years, the California price has remained quite stable. In addition, Union's statements prior to the plant's opening, such as the one about rising finding costs for crude, seemed to indicate the company took a very long view of the project and that yearly fluctuations in crude prices would not be a decisive factor in its decision-making, especially after mandatory import controls had been imposed. Regarding depletion, the disparity between shale oil and crude had existed for three decades prior to the plant's opening and it must certainly have been taken into account in Union's pre-opening estimates. There was no legislation pending at the time to close the gap.

Even after the plant was closed, Union officials indicated privately that their research had been successful. The BLM evidence contains a letter written by Albert C. Rubel on July 8, 1958, which states: ". . . (W)e can with complete assurance proceed with the building of a shale plant which we know will work . . . and at a price closely comparable to current crude prices." On February 27, 1959, Fred L. Hartley wrote that "As a result of our very large-scale shale demonstration plant and mining program, we have been able to establish firm costs. . . . (W)e believe that today's cost of producing shale oil is of the same order of magnitude as today's cost of finding and producing new sources of domestic crude oil."

In more recent public statements, Union draws somewhat different conclusions. John Pownall, who managed Union's plant, told the Oil Shale Advisory Board in 1964 that the company's studies indicated "oil shale was not yet sufficiently profitable to warrant the large-scale investment needed for a commercial plant." One of the biggest problems, he noted, was "assurance is needed that shale will face a stable economic environment."

In attempting to understand the complete reasons for the shut-down of the plant, some critics have pointed to a relationship between Union and Gulf Oil, one of the world's largest international petroleum producers with vast foreign reserve holdings. *The Black Bonanza* describes the beginning of the relationship this way:

Reese Taylor, then president of Union [until August 1956 when he became chairman], and W. K. Whiteford, president of Gulf, were long-time friends. From time to time they had discussed Gulf's excess crude oil reserves and production and Gulf's desire to sell large quantities of its production. Gulf had a surplus of cash. Taylor and Whiteford also discussed Union's need to finance its growing capital expenditures, as well as its long-term need for access to crude-oil reserves, particularly foreign reserves. Union and Gulf were noncompetitive.

During one meeting of the two presidents early in 1956, Taylor brought up the possibility of Gulf making a financial investment in Union. After further discussions Taylor suggested more specifically the sale of a convertible debenture issue by Union to Gulf. Whiteford expressed interest in the suggestion and negotiations between the companies began in March, 1956.

Upshot of the talks was the sale in April, 1956, by Union to Gulf of $120,000,000 3¼ per cent 25-year subordinate convertible debentures which Gulf might later convert into Union Oil common shares.

Conversion of these debentures, which were sold at a time Gulf was earning 13 percent on its total capital, would have given Gulf a 23 percent interest in Union, enough to control the company. It later became known that the debenture sale was the first step toward an eventual merger of the two companies.

Union began benefiting from the arrangement immediately. According to *The Black Bonanza,* Gulf signed a long-term contract with Union to supply it with oil from Gulf's huge Kuwait and Venezuela reserves (62 percent of Gulf's crude production in 1968 came from these two countries) and "guaranteed" Union enough crude to supply its fast-growing Western markets "until Union's exploration could catch up with the demand." In 1956, Union became a substantial importer for the first time, averaging initially 12,440 barrels/day. In addition to expanding its exploration program and retiring previous debt issues and preferred shares, Union used the new funds to order, for delivery in 1958, three new 60,000 ton "super tankers," each of which had a 415,000-barrel capacity. Until these ships were delivered, Gulf agreed to ship its crude to Union in Gulf's tankers. Eventually, Union's purchases from Gulf amounted to at least 10 percent of the company's refinery runs. Thus at the time Union shut down its shale plant partly because, it said, of rising imports, Union was a substantial beneficiary of the open import policy.

As soon as the merger plans between Gulf and Union became known, the Justice Department raised antitrust objections, at first forbidding Gulf to convert its debentures and finally forcing the two companies to call the plans off and Gulf to sell its debentures. In June, 1961, Union retired the Gulf issue by marketing another $120 million issue. The foreign crude agreement remains in effect.

On December 3, 1954, Dr. Tell Ertl, who since leaving Ohio State had become a well-known shale land speculator, wrote a letter to the chairman of the Oil Shale Advisory Board about Gulf's relations with Union. Ertl said the shale plant had been "successful beyond the highest anticipation of the protagonists" and that it had been closed down "at the height of its success, before complete results were obtained, and at a time when sales could have balanced out-of-pocket costs." Moreover, before Gulf made its deal with Union, he wrote:

. . . Mr. William Whiteford, the president and chairman of Gulf discussed oil shale with me several times, at least once in my home in Grand Valley, Colorado. During these discussions he and Fred Brandi, president of Dillon Read [Gulf's investment bankers] told me that Gulf's concessions in Kuwait terminate about 40 years from now, that the reserves in Kuwait appear undepletable before the termination of the concession at any foreseeable rate of production, consequently sound business is for Gulf to increase sales as much as possible. The inference left with me was that Gulf was willing to grant excellent terms to delay the inception of an oil shale industry.

The actual and circumstantial evidence is strong that Gulf did grant such terms to Union, did succeed in delaying the inception of an oil shale industry by making it appear more profitable for Union to accept Gulf's terms rather than continue the development of oil shale.

After I quoted Ertl's letter in a *Harper's* article on shale in August, 1968, Gulf Chairman E. D. Brockett, in a letter to *Harper's* editor-in-chief noted that "it has not been the policy of this corporation to attempt to answer every misrepresentation that appears in print concerning its history or its activities," but that for "the sake of accuracy" he wanted to "set the record straight" on Ertl's allegations concerning Gulf's activities. "Nothing," wrote Brockett, "could be further from the truth."

Ertl's shale land holdings give him an admitted interest in shale development and, moreover, he admits today that his "evidence" remains only "circumstantial," not "actual." It seems difficult, though not impossible, to believe that Gulf would have undertaken such a large step as a merger with Union only for the purpose of stifling oil shale development. Had it been pushed through Justice, the merger would have made excellent financial sense—the marriage of crude-poor with crude-rich companies is common in the oil industry. And at the time Union had extensively developed West Coast marketing facilities, which Gulf lacked.

It seems incontestable, though, that Union's deal with Gulf made producing shale oil less attractive or at least less compelling to Union, especially when a merger lay on the horizon that would have resulted in a combined company with all the foreign crude oil it would ever need. As the owner of huge reserves and as a producer of nearly twice as much oil as it refines, Gulf had nothing to gain from a shale oil production facility and might, if the facility had stimulated a shale oil industry, have had a lot to lose. Gulf's attitude toward shale research about this time is indicated by its refusal, in 1955, to join a project sponsored by the Denver Research Institute of the University of Denver which would have cost Gulf $15,000 a year over a three-year period. "It is our feeling," Gulf wrote the institute, "that in view of the minor interest that Gulf has in oil shale properties [Gulf had purchased a 3,760-acre

tract in 1952] a substantial support of this type of research is definitely premature. . . . We have already turned down other oil shale research projects of greater potential value." As the senior partner in the
contemplated merger—Gulf's sales in 1958 were almost seven times
as large as Union's—Gulf cannot have been expected exactly to encourage Union to continue its project. And in looking forward to a very bene-
ficial merger, Union may well have been hesitant about continuing with
a major, quite expensive research effort in an area in which its future
partner obviously had quite limited enthusiasm.

After 1961, Union's shale research resumed somewhat, and the company began working on a few small pilot retorts in its California research laboratories. But from 1960 to 1966, it spent only $390,000 on
shale research, about 1 percent of its total research budget. In 1964,
Harold E. Carver, a Union research associate, told the Oil Shale Symposium that except for the depletion allowance and import control
problems, "Union's complete technology and mining, retorting, and refining" was "ready for commercial application." Before the Interior
Committee in 1967, Fred L. Hartley announced that Union was building a new 140,000 barrel/day refinery near Chicago, and said "you will
be surprised and I hope pleased that we have incorporated into the design provision for processing 70,000 barrels per day of synthetic shale
oil." But he added that "as a precautionary measure and in recognition
of the failure of the executive branch of the Federal Government to establish a favorable and equitable climate for the oil shale industry, we
are also designing the plant to process 70,000 barrels per day of synthetic tar-sand oil from Alberta, Canada."

Hartley also announced that Union had just made an agreement permitting Battelle Development Corporation, a division of the nonprofit
Battelle Memorial Institute, to use the company's oil shale mine, buildings, and some 20,000 tons of crushed shale lying around in heaps for a
new research project. The idea had been put together by Battelle and
Development Engineering, Inc., a small Denver engineering firm owned
briefly by Battelle and now independent, for the purpose of testing the
feasibility of retorting oil shale with a 400 ton/day vertical kiln concept
designed by Development Engineering, a commercial version of which
was being used for lime production. The kiln would be constructed by
Koppers Company of Pittsburgh. In solicitations sent to several large
oil companies, Battelle urged them to join in a five-year program costing between $11 million and $12 million "to provide technical and eco-

nomic information to the participating companies to be used in their overall decision regarding shale-oil production."

A number of firms, including Shell, Sun, and Atlantic Richfield, expressed interest, but negotiations bogged down. One problem, according to Jack Bridges of Development Engineering, was that the oil companies couldn't agree on just how much of their own knowledge should be brought into the venture and how much credit they should get for prior research work. Another, says Warren Riley of Koppers, was that Battelle insisted that it retain ownership of all patents developed—Battelle derives much of its revenues from technology licensing. "The oil companies," he says, "weren't too happy about that." Both Riley and Dr. Joseph Oxley of Battelle claim the technological feasibility of the proposed apparatus was not a problem. Eventually discussions were terminated, and Union's plant today remains in "standby."

Halfway between Rifle and Grand Valley another narrow road leads from the main highway toward the Roan Cliffs to a small colony of white prefabricated houses and several tall complexes of pipes and girders. A 5½-mile switchback dirt road then leads tortuously up the cliffs 2,200 feet above the white buildings to several tall mines dug into the 73-foot-thick Mahogany Ledge shale layers. At one time, 300 men dynamited shale from the cliffs, which was transported down the dirt road in trucks, fed it into several retorters, and converted into many thousands of barrels of shale oil. Dozens of engineers worked in nearby laboratories studying the mechanics and chemistry of shale oil production. This facility, named Anvil Points after a nearby cliff formation, was built by the United States government and operated continuously from 1947 to 1956. Today it is closed down. Its only inhabitants are an eight-man caretaker staff.

Government-sponsored research into oil shale actually began during the height of the boom period of 1916–1920 and culminated in the construction by the Bureau of Mines of two experimental retorts outside Rulison, a now-deserted town not far from Rifle. One retort was modeled after the Scotch Pumpherston retort and another after the NTU retort, an intermittent, batch-type device that had been used to recover oil from a tar-sands-like formation near Casmalia, California. A supporting facility was set up in Boulder, Colorado, to study refining techniques. Both activities were modest and by 1929 had cost about $225,000.

Rulison was closed in 1929 and Boulder in 1932. No attempt was made to develop overall cost estimates.

After a period of quiescence, interest in shale revived during the Second World War when military usage began cutting deeply into domestic petroleum reserves. In 1944, the Synthetic Liquid Fuels Act, sponsored by Wyoming Senator Joseph O'Mahoney and West Virginia Congressman Jennings Randolph, was passed which directed the Secretary of the Interior through the Bureau of Mines "to construct, maintain, and operate one or more demonstration plants" to obtain "cost and engineering data for the development of a synthetic liquid-fuel industry." Possible sources to be investigated besides shale were coal and wood. Anvil Points was constructed on two "Naval Oil Shale Reserves" which along with the "Naval Oil Reserves" of Teapot Dome fame, were established by Congress in several actions beginning in 1911 to ensure that in time of war the Navy would have an adequate reserve supply of petroleum. The plant required an initial capital investment of $6 million and in the course of its research activities another $18 million was spent. It remains today the most thorough research effort ever made to find ways to economically extract oil from shale. It has also been the most controversial.

The chief controversy is over the circumstances of its closing in 1956, for it is undeniable that a group of high-ranking oil industry executives desired that the plant be closed, and that both the Interior Department and the Congress acceded to this desire. The oil industry first became directly involved with Anvil Points in 1950 when Interior Secretary Oscar L. Chapman requested an appraisal of the facility's cost data from the National Petroleum Council, an organization of senior oilmen set up after the war to advise the Interior Secretary. Meanwhile the Bureau of Mines made its own study.

Released in 1953, though based on 1951 data, the NPC report was almost unmitigatedly pessimistic in its conclusions: "As shown by this extensive and conclusive study, all methods of manufacturing synthetic liquid fuels [it analyzed coal as well as shale] proposed by the Bureau of Mines are definitely uneconomical under present conditions." It added that "the need for a synthetic liquid-fuel industry in this country is still in the distant future."

This contrasted sharply with the opinions of the Bureau of Mines and the Interior Department. Chapman stated in his 1951 annual report that:

Shale Economics Are Good

By virtue of these technical advances in mining, retorting, and refining, oil shale now occupies a very interesting economic position. The Bureau [of Mines] estimates that, with a capitalization of 50 percent equity and 50 percent borrowed funds and with all products selling at market values, the rate of return on the equity capital would be 11.2 percent after interest charges and income taxes. This return is based on an industry-scale operation and includes the cost of a pipeline to the West Coast. Although the return is less than the average return of the petroleum industry, as indicated by the published figures, it is high enough to warrant serious attention. Furthermore, oil-shale operation would not involve the exploration risks incident to crude oil production.

The report concluded that "The Department of the Interior believes that it would be prudent for private industry to establish a pioneer commercial oil-shale plant."

Curiously, despite these divergent summary appraisals, the mathematical conclusions of the two reports were similar. The Bureau of Mines' estimated return was much higher than that predicted by the NPC but most of the disparity lies in the different assumptions. The NPC, for instance, made its calculations on total equity capital while the Bureau of Mines figured only 50 percent equity.

Whatever the profit estimates, it was the opinion of Bureau of Mines officials that Anvil Points at the time was making marked technological progress. In mining, the goal of reducing mining costs in the "room-and-pillar" mines to 50¢/ton had been achieved and by 1954 a contemplated switch from percussion drilling to rotary drilling was expected to reduce the figure to 26.5¢. In retorting, experimentation had begun with a variety of ideas, but in the late 1940s, engineers had devised the Gas Combustion System, which was essentially an upside-down version of what was to be Union Oil's retort. Continuously operating and thermally self-sufficient, the apparatus was structured vertically so that crushed shale dumped in the top would move downward by gravity into a zone of combustion stimulated by an air vent. Heat was provided by ignited shale gas and coke. Like the Union retort, the shale vapors were condensed while moving through the shale rock, which they also helped heat, but unlike the Union idea, the shale oil was drawn off near the top. In both ideas all heat transference occurred in a single vessel. A 6 ton/day gas combustion retort went into operation in 1951, followed by a 25 ton/day model and, in 1953, a $600,000 150 ton/day version. The big retort ran into some immediate problems and was temporarily shut down pending experiments on the 25 ton/day retort. Meanwhile in refining, much work was being done on the elimination of impurities from the shale oil. Shale gasoline was used in cars and buses around Anvil Points, asphalts were used to resurface roads, and shale diesel fuel was used to power a variety of motors, including the locomotive of a Denver and Rio Grande Western Railroad passenger train during a trip from Salt Lake City to Denver.

Interior officials continued to be enthusiastic. On January 14, 1953, Under Secretary of the Interior Vernon D. Northrop wrote Senator John Sparkman that ". . . (T)he Synthetic Liquid Fuels Program is one of the most important research projects now being carried on by the government. . . . (I)t now appears that costs have been reduced to a point approaching a competitive level with petroleum products. . . . There is practically unanimous agreement among industries and Government that this program should be continued." President Eisenhower's new Interior Secretary, Douglas McKay, said at his nomination hearings in 1953 that he was "very strongly for" the synthetic fuels program. "I think we have to develop anything we can in the way of fuel these days," he remarked, "because we don't know when we will be at the end of our rope." A Bureau of Mines pamphlet published in March, 1954, noted that "Recent cost estimates are of interest because they show that the commercial cost of making shale gasoline and other liquid fuels on a large scale would closely approach the cost for the corresponding petroleum products."

Oil men continued to be unenthusiastic. As the industry's authoritative *Oil Daily* put it rather candidly in January, 1953:

The industry is willing to allow "pilot plant" experimentation to continue but only on a provisional basis, and only with control in its own possession. Even so some petroleum physicists and economists admit that within ten years, perhaps, and for some areas such as the Pacific Coast, the synthetic era may be nearer than most people realize. Technical progress may soon turn up the desired process. The industry is torn between the negative idea that synthetics aren't needed or are uneconomic, and the positive idea that if they become possible, they shall be exploited only by the present corporations.

Felix E. Wormser, the new Assistant Interior Secretary for Mineral Resources, meanwhile, appointed a five-man team including officials from the oil, coal, and copper industries, to make a survey of the Bureau of Mines' operations. In May, 1954, the team recommended, among other things, that:

The experimental work done solely by the Bureau on the production of oil shale and oil from shale at Rifle, Colorado, shall cease and . . . no further work be done with the new retort (heating mechanism) unless there is a substantial contribution by industry under a cooperative agreement. If industry feels that no further experimental work is necessary, then the facilities will have served the purpose for which they were developed and constructed, and disposition should be made in accordance with established procedures.

The team added that the Bureau should concentrate more on improving recovery of crude reserves and cooperating more fully with such industry groups as the American Petroleum Institute.

Secretary McKay then requested an opinion of this view from the National Petroleum Council. In January, 1955, an NPC "Committee on Shale Oil Policy" reported:

Various companies are now engaged in experiments and development work on processes for the recovery of oil from shale, and they and others will undoubtedly continue to carry on such work to the degree warranted by present and future circumstances. In the light of this fact, the Committee feels that there is no need for further government effort along these lines.

It recommended that work at Anvil Points should cease and that the facilities should be placed in standby condition.

Congressional appropriations committees went along with these recommendations and for fiscal 1956, though $1 million was appropriated for mining, no funds were allowed for retorting, and the retorts were shut down on June 30, 1955. For the following year, all money was cut off, and the entire facility was closed on June 30, 1956. *Oil-Shale Mining, Rifle, Colorado, 1944–56,* a Bureau of Mines report on the work at Anvil Points stated: "The Bureau of Mines developed a new retorting procedure that gave great promise for a more economical process. The pilot plant was operated for a few months before all retorting experiments ceased on June 30, 1955." The report described the new rotary drill idea which would more than double the amount of rock one man could drill and cut drill costs to one-fifth the former level. "The mine was closed before the research on rotary drilling was completed," the report said.

A number of Congressmen were highly angered by the shut-down. On March 1, 1955, Senator Estes Kefauver said on the Senate floor that "It is difficult to conceive of a more clear-cut case of oil-company domination of the policy of the United States Government." He quoted from a 1954 *Denver Post* editorial which said:

If a jury of railroad presidents was asked to decide whether the trucking industry should be allowed to use public highways, there would not be much doubt what the verdict would be. We do not leave decisions on whether the Government should subsidize the airlines to committees composed of bus-company executives. . . . We expect our government, in passing on any matter involving Government authority, to act in the best interests of the people, not to line up with any group of competitors against another. But Secretary of the Interior McKay, in an almost unbelievable action, has asked the liquid-petroleum industry, as represented by the National Petroleum Council, to advise the Government on whether Federal appropriations for oil-shale experimental work should be stopped. . . . The oil industry is just about as anxious to have competition from shale oil as the Republicans are to have the Democrats win the November election.

At hearings before the Senate Appropriations Committee, Republican Senator Eugene D. Millikin of Colorado, who has an admitted interest in a few tracts of shale land, said:

I have been in this oil shale business since 1919. I know of my personal knowledge who has done work in the field and who has not. There is not one on that Council unless it would be the Union Oil Company, that has made any contribution toward the oil shale business, which is different from the petroleum business. I want to be very conservative, but it was a very nervy performance for people who were in the liquid petroleum business to come along and suggest that this potential competitor of that industry be put out of business and that the plant be mothballed. I think it was misconceived.

Also unhappy were several members of the Navy Department which, as custodian of the Naval Oil Shale Reserves, has maintained a continuing interest in the status of shale research. Captain Albert S. Miller, Director of Naval Petroleum, testified before the Senate Appropriations Committee that:

One large factor contributing to [the NPC's statement that private industry was undertaking sufficient oil shale research] is the recent announcement by Union Oil Company to build a $5 million plant to process oil shale. What the committee did not know was that this pilot plant is being built to find the answers necessary to construct a commercial plant. Should the answers not be found within a 2-year program, the company will discontinue the project. In that case the government stands to lose the money which has been spent at Rifle and the key to unlock billions of barrels of oil has not been found. Should the company be successful in its efforts, then all patents and processes belong to the company. Should the government continue its efforts and obtain success, then all patents and processes are in the public domain.

With the exception of the minor effort by Union Oil Company, what major segment of the oil industry has rushed in to provide funds or set up a program?

In 1957, the Navy submitted to Congress detailed plans based on a study by Koppers Company for a research program which, though financed by the government, would be contracted out by the Navy to private industry. The idea, Miller told the House Armed Services Committee, would be to perfect an economical shale oil process with a large-scale 1,000 ton/day retort. Then if petroleum shortages should develop during wartime, the Navy could simply take the plant "off the shelf," put it into operation, and build duplicate plants if necessary to solve the fuel problem. He pointed out that despite possession by Standard Oil of New Jersey of patents and technology on synthetic rubber processes before World War II, after war had broken out and created drastic rubber shortages, a $700 million crash government program was required to underwrite a synthetic rubber industry. (The plants were sold to the major oil and rubber companies after the war.) Miller today claims that he had Representative Carl Vinson, head of the Committee, sold on the idea: "I was keeping Uncle Carl very well informed." After all, a special Subcommittee on Petroleum of Vinson's committee had already reported in 1948 that the government should "expedite research" into synthetic fuels development, though "without impeding presently used processes of producing our national petroleum."

The Navy's plan ran into hard opposition. "The Department of Interior fought like mad," says Warren Riley of Koppers. "They felt that oil shale was their purview, and that if anybody got money for oil shale it ought to be them." The oil industry was also very opposed. "The last thing they wanted to do was to encourage a competitive oil shale industry," says Jack Bridges of Development Engineering, Inc., who was Miller's assistant and spent two years at Anvil Points developing the proposal. Miller felt that the oil industry was especially fearful that the Navy wanted to get into the oil business itself—indeed Navy Secretary Josephus Daniels had urged just such a move in 1913—and he worked assiduously to assuage these feelings. "I would like to emphasize once more," Miller told the Colorado Mining Association, "that the Navy has never had and never will have any intention of competing with private enterprise in the development of an oil shale industry." Development instead will depend, he said, "upon industry and government going hand in hand in a spirit of cooperation."

His efforts were in vain, for no action was ever taken on the proposal. It is likely, in fact, that the Rifle plant would have been sold or torn down had it not been for what Miller called "the fast action of certain Congressmen," principally Colorado Representative Wayne N. Aspinall, whose district includes the Piceance Creek Basin. Aspinall helped push through legislation allowing the Navy to use part of its general appropriation for the Naval Petroleum Reserves for the maintenance and custody of Anvil Points. "When you get through all the speeches and all the talk down to the nitty-gritty," says Jack Bridges, "99.98 percent of all the progress that's been made on oil shale has been made by Wayne Aspinall."

Aspinall's concept of progress is not always synonymous with that of other oil shalers. He actually favored the closing of Anvil Points, he says, because it "wasn't getting anywhere" and had become a "sanctuary of bureaucratic administrators." A vocal advocate of the oil industry's depletion allowance, import controls, and other financial benefits—his district is a major crude oil and gas producer—he is also a hardy opponent of federal research on oil shale. "I don't think the government should do that kind of work," he maintains, "and start trying to retort and develop the shale and so forth." After several years of squabbling between Interior and the Navy over just who should control Anvil Points, plus argument with the Attorney General who had ruled neither the Navy nor the Interior Department had the authority to lease the plant to private industry, Aspinall finally managed to get legislation passed in 1962 which authorized the Secretary of the Interior to take possession of the facility and, after approval of the President, to "contract, lease, or otherwise encourage the use of the facility . . . in research, development, test, evaluation, and demonstration work." A section of the law stated that nothing "shall be construed . . . to authorize commercial development and operation of the Naval Oil Shale Reserves by the Government in competition with private industry."

Following approval of the legislation, Interior asked for proposals from parties interested in using the plant, and got a response only from Garrett Research and Development Company and Mobil Oil. Garrett subsequently withdrew, and an agreement was signed and approved by President Johnson in April 29, 1964, to lease the facility to the nonprofit Colorado School of Mines Research Foundation, which is closely allied with the Colorado School of Mines and receives about 98 percent of its annual income from private industry. The Foundation then immediately subleased it to Mobil and Humble Oil, who also had joined, for a period of five years with an option to renew for another five years. According to Dr. Orlo Childs, who heads both the School and the Foundation, "Interior was a little leery of giving it to Mobil directly and then hearing the hue and cry of giveaway. We're a kind of buffer, so there won't be all the ramifications on the Washington scene that there would be if industry got it directly."

Other "qualified" parties were invited to join Mobil, who was to be "Project Manager," and Humble within six months. Eventually, Continental Oil, Pan-American Petroleum Company (a subsidiary of Standard Oil of Indiana), Sinclair Research, Inc. (a subsidiary of Sinclair Oil), and Phillips Petroleum came in. "It took a real salesmanship job to get them in," admits Aspinall. "There are an awful lot of people who aren't interested in anything but a holding action on oil shale."

There may have been a variety of reasons for participation. In a deposition for the Bureau of Land Management Contests, Robert H. Cramer, Mobil's program manager at Anvil Points, said that:

Mobil felt that of the retorting processes under investigation, that the gas combustion appeared the most feasible. Mobil also felt that of the mining methods under investigation the room-and-pillar appeared the most feasible. Therefore, it was a very logical extension to attempt to pick up the work on the gas combustion retort and the room-and-pillar method which had been done by the Bureau of Mines at Anvil Points and carry it further. . . . Furthermore, the facilities were in being and had been maintained in reasonably good shape during the eight years that they were shut down, and it was much faster and much more economical to move into an already-built facility than try to go out and build a facility of our own.

The BLM evidence quotes a letter from the Denver area vice-president of Shell Oil to the New York office in August, 1963, requesting permission to:

send a letter of intent to the Colorado School of Mines Research Foundation, Inc., supporting their proposal for operation of the Rifle facilities (see attached brochure). This is basically a maneuver to keep the plant in neutral hands, to provide an additional avenue for keeping abreast of technical developments and to gain a certain amount of good will with Colorado officials in state and national government.

The New York office, however, decided against joining.

Those who did join were probably pleased with the official agreement, a 114-page document a team of 30 lawyers spent six months hammering out. Especially interesting were the provisions pertaining to ownership of rights to technology. Article III of the initial "Lease Agreement" between the government and the Colorado School of Mines Research Foundation stated that: "Research Foundation agrees to assign to the Government, as represented for this purpose by the Secretary of the Interior, all its right, title and interest in and to all Subject Inventions." A Subject Invention is, basically, an item of technology which may be developed and for which a patent may be applied. Thus, the provision meant that the government owned the patents. However, "Appendix I" to the Lease Agreement, agreed to by the government, the Research Foundation, and the oil companies, superseded Article III and stated that "Any and all Inventions (and all applications for patent and patents thereon) . . . shall be the property of the Parties [i.e., members of the consortium] jointly." The government had to be satisfied with "irrevocable, non-transferable, royalty-free" patent clearances, or the right to use them, "for operations and activities by or on behalf of the Government." Between 3 and 20 years after the termination of the program, the consortium members agreed to grant licenses on their patents to "responsible applicants" at a "reasonable royalty."

Appendix I would appear to depend on a rather liberal interpretation of the Government Patent Policy as promulgated on October 10, 1963, which "seeks to protect the public interest by encouraging the Government to acquire the principal rights to inventions in situations where the nature of the work to be undertaken [under grants to private industry] or the Government's past investment in the field of work favors full public access to resulting inventions." It states that the government shall acquire exclusive rights to inventions, among other situations, where "the contract is in a field of science or technology in which there has been little significant experience outside of work funded by the government, or where the government has been the principal developer of the field, and the acquisition of exclusive rights at the time of contracting might confer on the contractor a preferred or dominant position; or the services of the contractor are for the operation of a government-owned research or production facility. . . ." The only case where private parties can acquire the rights is when the work being performed "is in a field of technology in which the contractor has acquired technical experience (demonstrated by factors such as know-how, experience, and patent position) directly related to an area in which the contractor has an established nongovernmental commercial position." The Secretary of the Interior justified acceptance of Appendix I in a "Certificate" which said that "substantial proprietary data and experience [of the participants] will be drawn upon in the research and experimentation," and that "the Parties have agreed to bear the cost. . . ." Interior lawyers contend they rather sharply limited the scope of the project to the already existing mine and retorts—the oil companies had wanted to conduct *in situ* research, for example—for which the government already holds patents.

Another noteworthy provision of the lease agreement gave the government access to all data and information generated, but said it must not release publicly such material until three years after completion of the research. The participants were not required to divulge their interpretation or evaluation of the data, and in their annual report to the Secretary of the Interior they were not required to submit anything they regarded as "confidential."

In other words, eight years after the oil industry had suggested that government operation of Anvil Points be terminated, a closed-end (after six months) consortium of six major oil companies, none of whom had ever conducted any appreciable research on oil shale, let alone established a "commercial position," was being given complete access to $6 million worth of publicly owned equipment in which $18 million in public funds had been invested and which constituted the most ambitious shale research facility ever developed. The consortium was being given ownership of all technology that might be developed. The government could only release information generated by the project—that is information not withheld by the participants on the grounds it was "confidential"—a maximum, if the lease were extended, of thirteen years after the program had begun. J. Wade Watkins, Director of Petroleum Research for the Bureau of Mines, told the Senate Antitrust Subcommittee in 1967 that "there was some concession made by the Government patent attorneys in this particular agreement because of the fact that the Government was putting no money into operation of the project."

A number of Justice Department antitrust lawyers were very concerned about the concession. Just a few days before it was supposed to be signed, a draft of the agreement was submitted to Justice, who swiftly raised several objections. It appeared that it might take a considerable amount of time to work something out. A man particularly upset by the possibility of a delay was Congressman Wayne N. Aspinall who had planned to make a dramatic announcement of the signing during his speech scheduled for May 1, 1964, before the Colorado School of Mines' First Symposium on Oil Shale. After some heated discussions, Justice agreed to a compromise: the department would approve the agreement if President Johnson's letter authorizing the lease contained a clause indicating that nothing in the agreement freed the participants from prosecution under the antitrust laws. In a letter to Senator Philip A. Hart, chairman of the Senate Antitrust Subcommittee, on May 8, 1964, Assistant Attorney General William H. Orrick, Jr., explained the compromise as follows:

It appeared to us that the proposals posed several possible antitrust problems. These, however, could not be finally resolved prior to the execution of the agreements, since resolution depended largely on the course of conduct followed by the participants, the whole context of petroleum industry research, and the practical effect on other competitors of the arrangement. We therefore suggested that, if it were considered otherwise administratively desirable that these arrangements be consummated, antitrust's prime concern would be protected if care were taken to avoid any appearance that the participants were immunized from any appropriate antitrust enforcement action trying these issues subsequently after due investigation.

President Johnson's letter, containing the appropriate clauses stating that nothing in the agreement shall relieve anyone from "the operation of the antitrust laws," came through right on schedule: April 29, just two days before Aspinall's speech.

In February, 1966, the members decided against proceeding any further and announced that their program had been "completed." As of April 13, they relinquished the plant to the government. During the project's three years and nine months, the six parties said they spent a total of $7.2 million, about $2.2 million more than they had originally planned. For Mobil and Humble, who each spent about $30 million annually on research, Anvil Points ate up about 1 percent of their budgets. The same week Anvil Points was reclosed, 25 oil companies paid $603 million for lease rights to 363,000 acres of federal land on the Outer Continental Shelf in the Santa Barbara Channel.

Whether the reopening of Anvil Points was a great leap forward toward commercial shale oil production is difficult to say. The first public release of data from the experiments was a report by the Bureau of Mines in November, 1969, based on data and technical information made available by the oil companies to two BuMines "observers." The report contains a wealth of statistics, but makes no conclusions about cost, commercial feasibility or even whether or not shale technology had been advanced beyond that developed by BuMines in the 1950s.

Several explanations for the consortium's unwillingness to continue work besides a possible lack of technological progress are conceivable.

A Mobil Oil executive points out that Justice had been starting on an ambitious antitrust investigation aimed at the oil industry's propensity for joint ventures. "A company counting on a joint venture has to worry about that problem," he says. Warren Riley of Koppers Company says on the basis of conversations with participants that the members' ambitions were not too high to begin with. "There was a great deal of reluctance to come up with any really new ideas," he contends. "The oil companies weren't really interested in developing something that would result in patents to which the government had free access."

Dr. M. L. Sharrah, Continental Oil's vice-president in charge of research and engineering, told me in 1967 that on the basis of his company's work at Anvil Points up to that time, he felt oil shale costs to be "damn near exactly competitive" with the cost of finding and producing crude oil. His company's purpose at Anvil Points, though, was merely to work out the design for a prototype plant "and then we'll put it on the shelf and wait for the competitive bidding" on the government shale land. "In order to bid intelligently," he went on, "you have to know as much as possible."

In June, 1968, the Bureau of Mines announced that while it had no plans of its own to conduct further research at Anvil Points it would consider proposals by interested parties for use of the facility—anything, it said, from short-term lease to outright sale—and it would then select the offer that "would best serve the public interest in advancing research and development on the production of oil from shale." By the July 31 deadline, only a single proposal had been received—from Development Engineering, Inc., the Denver firm whose liaison with Battelle Development Corporation had proved unfruitful, who said they wanted to conduct experiments with some of the facility's equipment. "Their proposal is sort of hanging fire at the moment," said a Bureau of Mines official in November, 1969. "We don't know when we'll do anything on it." Wayne Aspinall appears to evince a certain frustration with the rather jejune trend of events. "I've served notice on the government," he grumbles, "that unless they put it back to use that they should have it dismantled."

Except for its caretaker staff, meanwhile, Anvil Points remains deserted.

The most alluring method, at least in theory, for producing shale oil is *in situ*. Immense benefits would ensue if oil could be obtained from the shale while it was still in the ground. Mining, which creates the bulk of normal shale oil production expenses, could be eliminated. The huge capital costs and lengthy lead times would be unnecessary. The problem of waste disposal would be nonexistent. The surface of the land would remain unscarred. There would be no air pollution. A number of oil companies have been somewhat more interested in *in situ* than surface retorting, and have taken the trouble to file patents on their ideas, for while the rim of the Piceance Creek Basin will probably be mined and retorted conventionally, the thick, rich layers of shale in the government land, covered by extensive overburden, may be especially amenable to *in situ*, if and when the government gets around to leasing it.

Despite a large number of theoretical studies, actual testing of *in situ* ideas has been rather scanty. The principal technical problem is that most oil shale beds have practically no permeability or porosity and they are rather poor conductors of heat. Some investigations have been made into the possibility of injecting hot gas or some other heating substance into wells, and recovering retorted shale oil in nearby producing wells interconnected with the injection wells. In Sweden some success was obtained during the 1940s by drilling holes into rather shallow shale beds and then heating the formation with large electrical resistance elements. But most shale experts agree that the most feasible and economical method involves fracturing the shale bed into small pieces before commencing retorting. During 1953 and 1954, Sinclair Research Corporation tested fracturing by hydraulic techniques, then igniting the rubble with propane gas and gasoline. Once burning, the combustion zone was fueled by coke and shale gas, the retorting by-products, and air pressure into the fracturing area was used to push combustion through the shale. Shale oil was driven off through holes drilled into the fracture area and pumped to the surface. Other varieties of this concept have been tried or postulated. Hot natural gas, superheated steam, explosives, and electricity have been the principal fracturing agents suggested though most actual experiments have been conducted in naturally fractured shale. There are also many ideas on ways to control combustion through air and gas flow. The only presently operating government oil shale research effort, at the Laramie Petroleum Research Center in Wyoming, has been focusing its work on *in situ*. The project though is very modest. A report in June, 1969, discussed the use of nitroglycerin, hydraulic pressure, and electrolinking to fracture shallow, 20-foot beds of shale near Rock Springs. Propane gas was used for ignition, and once begun, combustion was sustained only by air pressure, the report said. Maximum achieved production, though, was only 4½ barrels/day, and no estimates were made of efficiency or economics.

Over the past decade, much thought has been given to the possibility of using an atomic bomb to fracture the shale. Studies of underground nuclear explosions in other types of rock formations over the past 13 years generally reveal this chain of events: Within a few microseconds after the blast, temperatures up to 10 million degrees vaporize and melt the nearby rock, and the great pressure creates a fairly large cavity while fracturing the rock further away. Any time from a few minutes to a few hours later, the roof of the cavity begins falling in, eventually creating a "chimney," a vertical column of crushed rock perhaps hundreds of feet high. The chimney, the theory goes, would represent a huge NTU, batch-type retort. A hole would be drilled into the top, the rubble ignited, and combustion pushed downward by compressed air. Shale oil would dribble to the bottom of the chimney and be pumped to the surface. The center of the Piceance Basin would be ideal for nuclear *in situ,* its advocates point out, because the thick shale beds would permit very large chimneys and the heavy overburden would prevent escape of radioactivity.

Discussion of nuclear *in situ* began in 1958 among engineers of the Atomic Energy Commission and the Bureau of Mines. Both groups sponsored a meeting in Dallas during July, 1959, and proposed the idea to about 200 oil industry executives and research personnel. As part of the AEC's Plowshare program begun in 1957 to explore the peaceful uses of atomic energy, the project, the AEC said, would entail the government putting up $1 million for an atomic "device" (the word "bomb" is steadfastly avoided by Plowshare personnel) plus half of the estimated $1.2 million cost of the rest of the experiment. Private industry would contribute the remaining $600,000. If everyone agreed, the AEC went on, there could be a shale shot as soon as 1960.

Private industry was unimpressed, however. The AEC spokesman had mentioned $1 a barrel as a possible cost, but had stressed that until a shot actually went off it would be impossible to evaluate commercial feasibility with any accuracy. Oil executives said they wanted firmer information, and until it was forthcoming, they maintained they did not want to put up any money.

No further progress was made until January, 1965, when two firms who had done research work for the AEC—Edgerton, Germeshausen & Grier, Inc. (now known as E.G.&G.), and Reynolds Electrical & Engineering Company, Inc.—joined with Continental Oil to create CER Geonuclear Corporation with the expressed purpose of developing industrial applications of nuclear explosives. Continental Oil's special interest was stimulation of underground natural gas deposits whose production was impossible by present methods. CER solicited all of the major oil companies, as well as companies in other fields, on their interest in a nuclear blast in oil shale, and after a July, 1966, meeting 26 corporations agreed to participate in a research study which entailed individual annual contributions of about $40,000. Eight companies later dropped out. Of the remaining firms, 13 were large oil companies: Atlantic Richfield, Cities Service, Continental Oil, Getty Oil, Marathon Oil, Mobil Oil, Shell Oil, Sinclair Oil, Sohio Petroleum (a subsidiary of Standard Oil of Ohio), Sun Oil, Tennaco Oil, Texaco, and Superior Oil. The others were Equity Oil Company, a small independent oil company with large holdings of shale land; Western Oil Shale Corporation, an owner of tracts of Utah shale land; El Paso Natural Gas, a participant in other CER projects; The Cleveland Cliffs Iron Company, a mining company involved with The Oil Shale Corporation in a research project described in detail in the next chapter; and the Union Pacific Railroad, which owns some shale land. Gulf Oil was one of the big oil companies which declined to participate. The BLM Contest evidence quotes an internal memorandum on the July, 1966, meeting by R. J. Wyllie, Gulf's representative:

In view of Gulf's balanced crude position at home and abroad, and its in-house expenditures on production research, there is no reason to spend money on a joint shale-oil project. The project itself is . . . far too much of a long shot for any but crude-short companies even to consider. We will continue to keep an eye on the progress of the proposed scheme. Beyond this, I propose to take no further action.

Just three months before the consortium was formed, something of a sensation was being created by Mitchell A. Lekas, then a project engineer with the AEC, at the Third Annual Oil Shale Symposium. These symposiums are traditionally attended by oil shale's most fervid proponents, for whom good news during the past ten years has been as rare as a barrel of shale oil. As the 400 members of the audience went into what may well have been mass cardiac arrest, Lekas announced that potentially oil could be produced from shale by nuclear explosives for a mere 29¢ a barrel, a cost Lekas termed "surprisingly low," which, all things considered, was something of an understatement.

Shale enthusiasts and promoters of those small companies whose assets consist mainly of shale land were later to maintain or at least imply that this estimate meant that *all* of the shale could almost *definitely* be produced *right away* for that price. Speculators in shale land firms swiftly subtracted 29¢ from $3.00, multiplied the result times the number of barrels conceivably contained in the land owned by their companies, and, swept along by their euphoria, arrived at corporate assets and ultimate profits that all but dwarfed General Motors. Shale stocks jumped handsomely.

Though Lekas' estimate was revealed upon more careful examination to be very hedged, it certainly did sound encouraging. Costs, he said, moved downward as the thickness and the richness of the shale increased. Assuming 75 percent recovery, a shot in 15 gallon/ton shale beds 200 feet thick might produce shale oil for $2.04 a barrel, including capital costs but excluding land costs, treatment of shale oil prior to the refinery, and transportation charges. The magic 29¢ was obtained, he said, only under optimum conditions: 25 gallon/ton shale in beds 1,000 feet thick. There are, however, some 670 billion barrels of oil contained in beds 1,000 feet or more thick. Lekas sketched details of a possible industrial plant. On 1,000-foot shale, there would be four underground chimney complexes each created by five 500-kiloton bombs over a 400-acre area. This plant, he said, would yield 75,000 barrels/day or about 214 million barrels over an eight-year plant life. In questioning after his speech, Lekas stressed that his figures were very rough but that they were in the "ball park." "I don't think we'll be too far away," he added. At the same symposium a CER official said he believed the first shale shot would come in 1967.

It was not until October, 1967, though, that CER, the AEC, and the Bureau of Mines produced their research report, of which Lekas was a coauthor, on what was to be called "Project Bronco." It recommended that a 50-kiloton shot be exploded 3,350 feet below the surface in a plot of federal land about 38 miles northwest of Rifle in the heart of the Piceance Basin. It outlined a five-year program, with the shot coming about 10 months after the go-ahead. The blast, it said, would create a chimney 520 feet high, 230 feet in diameter, containing over a million tons of fragmented rock. A *Wall Street Journal* story quoted estimates of knowledgeable officials that $4 million would be spent by the government and $7 million by private industry. The report made no per barrel cost predictions but said that "The technology of fragmenting and fracturing rocks with nuclear explosives is well developed. The technology of retorting oil shale in aboveground retorts has been under investigation for many years. This study concludes that these two technologies could be combined into an economically attractive industrial process. . . ." A CER official told me in 1967 that a single 150-kiloton bomb could fracture shale containing 45 million barrels of oil. He estimated final recovery costs of as low as 60¢ /barrel. Since large bombs are only slightly more expensive than small ones, costs would presumably drop as the operation became larger. Meanwhile, in a paper that year for the Seventh World Petroleum Congress, some CER executives predicted that the test shot "should begin in 1968 with the *in situ* recovery starting in 1969." In April, 1968, the *Wall Street Journal* quoted H.H. Aronson, a CER vice-president, as saying the first shot would be "the summer of 1969."

Two other Plowshare projects were at the time moving swiftly toward the explosion stage. Project Gasbuggy set off a 26-kiloton device on December 10, 1967, near Farmington, New Mexico, to test the stimulation of natural gas recovery from dense underground reservoirs. The industrial participant with the AEC and the Bureau of Mines was El Paso Natural Gas.

Project Rulison, a joint effort by the AEC, Bureau of Mines, CER, and Austral Oil Company, a small Houston firm, also designed to stimulate natural gas, was set off on September 10, 1969, in the rich Mesa Verde gas fields near Rifle, just across the Colorado River from the Piceance Basin oil shale lands. The 40-kiloton shot occurred just two years after it had been first proposed to the AEC by Austral. Though actual commercial exploitation of nuclear explosions would not be possible without a Congressional amendment to the 1954 Atomic Energy Act, *Business Week* reported in October, 1968, that "U.S. companies have suddenly caught fire over the possibilities of using underground nuclear explosions to release huge quantities of natural gas and oil—as well as some minerals—locked in vast deposits beneath the surface of the earth." However, at a symposium on Project Plowshare in January, 1970, a CER official admitted that "costs of the [gas stimulation] operation must be reduced drastically or it will never be economic" and that "radiation has emerged as the major problem to be solved." Three months later, CER more optimistically announced a $300,000 contract from Equity Oil to study gas stimulation in the Piceance Basin.

Project Bronco appears to be facing almost as bleak a future as the town of Rulison, an oil shale ghost town near Rifle after which Project Rulison was named. So far the government and the CER consortium have been unable to agree on the contract terms. In the first draft submitted by CER and the consortium to the government in December, 1966, the consortium members wanted, among other things, exclusive rights to all technology and patents that came out of the blast, though during the nuclear fracturing phase it gave the AEC the right to file for patents if the AEC desired. The consortium also wanted a "closed patent pool" that would prohibit outsiders from participating. J. Wade Watkins, Director of Petroleum Research for the Bureau of Mines, told the Senate Antitrust Subcommittee in 1967 that in the draft there were some "antitrust implications arising principally from patent pooling arrangements, restrictions on disclosure of information, and restrictions on uses of patents by licensees. . . ." George Miron, Associate Solicitor of the Interior Department and a representative at the negotiations, said that CER was told the closed patent pool would be "a violation of the Sherman Antitrust Act." Also, Miron says, the government was disturbed by provisions restricting release of information until three years after the end of the project, which was planned for five years. "That would have given them an eight-year lead time on everyone else," he claims. "The government just couldn't agree with anything like that."

After lengthy negotiations between CER officials and government lawyers, another tentative agreement was reached in July, 1968. But when the draft was circulated among the consortium members, virtually all of the major oil companies refused to accept it, for the government had insisted on a number of substantial charges. Now they would have to relinquish to the government patent rights for all phases of the project including retorting. The government would then be able to make them available to outsiders. The government's participation in planning and execution of the project was increased, and disclosure rules had been loosened. And following a feasibility study, the cost of the project had risen from something over $6 million to as high as $18 million. While the first draft had permitted the parties to withdraw at several points before the project termination, the AEC wanted the members committed to the full duration.

As of the beginning of 1970, there had been no resumption in negotiations, and CER officials said they have had no contact with the consortium members for "over a year." J. Wade Watkins says the oil industry's interest "cooled off" after the North Slope discovery and that "I guess you can say Bronco is in a state of limbo."

A survey of consortium members for this book showed that most regarded Bronco as being closer to a state of expiration if not complete interment. A Standard Oil of Ohio executive says it has "fallen apart." Another oil man says it has "collapsed." Virtually everyone talks about it in the past tense. This survey, plus interviews with other participants, indicates perhaps five reasons for the inability of the consortium and the government to reach an agreement:

1) Inevitably, the logistical problems of getting eighteen teams of lawyers from companies with diverse interests to agree on anything as potentially meaningful as oil shale development are immense.

2) Public concern has been increasing over the possible detrimental effects of underground nuclear explosions such as earthquakes, tidal waves, water pollution, and release of radiation. Many protests were raised against Rulison, and the American Civil Liberties Union even tried, unsuccessfully, to obtain a court restraining order. Outcries against a military blast in the Aleutians in September, 1969, were even louder.

3) Though a $1 million outlay per company spread over five years seems hardly an unreasonable expense for what CER called in its 1967 report a possible "major breakthrough" in shale, most of the consortium members were very critical of the cost. And they didn't like being committed from the outset. "That's no way to run a partnership," the Standard Oil of Ohio man complained. "They just didn't have any faith in us. We wanted to be free to leave if things didn't work out, and if it's going to cost that much money, well it probably wouldn't have been economical anyway."

4) There appears to be widespread doubt among the consortium members as well as many independent engineers that the project has much chance of working. Because oil shale is so very impermeable and nonporous, as well as inelastic and tough, the fractured rock chunks may be much too large, making heat transfer haphazard and efficiency of recovery low. Even if the rock chunks are of a reasonable size, it is very uncertain how uniformly the combustion zone can be moved through the nuclear chimney. Some people feel the eventual percentage of the oil in the fractured shale will be closer to 5 percent than Lekas' 75 percent, which would mean fantastic resource waste. Extraction of the other minerals contained in the oil shale, such as sodium and alumina, might be impossible.

AEC and CER engineers, nevertheless remain optimistic and contend their knowledge of fracture patterns from over 200 underground shots since 1957 makes them almost certain the chimney will form as desired. Retorting, meanwhile, is relatively simple, they claim. Mitchell A. Lekas, who left the AEC to form a company called Geokinetics, Inc., says that during 1967 and 1968 a research team at the Laramie Petroleum Research Center, for which he was the AEC's representative, simulated in a 10 ton/day aboveground retort the conditions they felt would exist in a nuclear chimney. Shattered shale, consisting of a wide variety of particle sizes, some a foot thick, was dumped into a cylinder 6 feet in diameter, 12 feet high, and then ignited. "We found we could easily control the movement of the combustion front," Lekas asserts, "and we got recovery efficiencies as high as 80 percent. The very least you would get would be 50 percent. I'm still very bullish on the idea." (Laramie recently completed "highly successful" first tests with a 45-foot-high, 175-ton retort which produced a 60 percent yield.) Other engineers feel that recovery of associated minerals will be possible through water leaching techniques, which will dissolve out sodium minerals after the shale has been retorted. They admit the idea is very theoretical, however.

In interviews, AEC, CER, and Bureau of Mines people typically after five minutes or so of theoretical discussion begin displaying considerable frustration and say something like, "How the hell can I give you a positive answer to that question until we set one of the damn devices off?" "Let's get on with the show!" two CER scientists told the Third Oil Shale Symposium. "We will never know how successfully we can retort oil shale *in situ* until we give it a try. . . ." The oil companies, apparently, did not agree.

5) The chief roadblock, as in so many other shale research ventures, was control of the technology and the patents. The confrontation on this point, between the Interior Department and the oil industry, was irreconcilable, as it has been in so many other areas. Interior, fearing criticism and anxious to uphold its version of "the public interest," was unyielding and stern, though to a large degree it had the weight of law and precedent on its side. The oil industry, though, was understandably reluctant to help finance and encourage oil shale technology applicable mainly to the public land, which would be available to anyone and which might, just might, yield something close to the magic 29¢. They would simply have no contractual control over what might happen. "The government's restrictions were just unreasonable," said a Standard Oil of Indiana official. "Those parts about giving third parties access just didn't make sense." "The contract terms were unacceptable," said

the executive of another oil company who declined to allow his company's name to be used, "especially the government's far-reaching demands about opening up research results."

6) The oil industry's aspirations for Bronco may not have been especially lofty to begin with. "I was really surprised to see all those companies in there," said Mitchell Lekas. "Of course the real interest of oil companies in shale is inversely proportional to their crude reserves, and the ones in there with a lot of reserves are only trying to protect themselves." George Miron, who has left Interior to join a Washington law firm, commented, "The whole thing was hustled up by CER, and the oil companies were afraid that if they didn't get in, CER would go ahead without them. If something was going to happen, they felt they better be there, especially if some of the others were there."

One of the members of the consortium, though, is working hard to stop the Bronco idea from dying. Western Oil Shale Corporation recently hired CER to make some investigations of Western's shale lands near the center of the Uinta Basin in Utah, about 120 miles southeast of Salt Lake City. The land is part of a 76,000-acre plot the company holds on a 20-year lease from the State of Utah for which it only pays 50¢ a year in rentals.* The land is much like the Piceance Basin— 3,000 feet of overburden covering a thick layer of rich shale—and Western hoped to have made a detailed feasibility proposal to the AEC by early 1970 for "Operation Utah." If the proposal is approved, said Ted B. LaCaff, Jr., until recently president of Western and of Texas American Oil Corporation, a small independent producer of oil, gas, and other minerals which owns 55 percent of Western, the company will invite four or five "majors" who were members of Bronco to join Operation Utah as "sponsors."

One of Operation Utah's attractions, LaCaff pointed out, is that since it will occur on private land instead of public land, the Interior Department will not have to be involved in the negotiations. Interior, of course, is the parent agency of the Bureau of Mines that will be in-

---

* Utah is the only one of the three shale states to adopt a leasing policy of state-owned shale lands. All of the states, especially Colorado, have passed other legislation to encourage a shale industry. Colorado has granted potential shale producers a number of attractive state tax breaks. Beside Western Oil Shale, three major oil companies—Shell, Pan American Petroleum and Atlantic Richfield— also hold sizable leases of Utah shale land. The state imposes no time requirements for development of its land, and though the oil companies obtained their leases during 1963–4, none has conducted anything more than exploratory core drilling. Nearly 223,000 acres of Utah shale land is currently under lease.

volved. However he felt that the change in administration in Washington is a hopeful sign. While his opinion of Interior lawyers under Stewart Udall was unprintable, he was more favorably disposed toward the new Solicitor, Mitchell Melich, a former Utah mining lawyer. Besides, he added, "we've got the governor, the senators, and the representatives right behind us. Once the government realizes how they blew the last deal by hanging tough, they'll be a lot more cooperative." His prediction as of December, 1969, was a nuclear shot by late 1970.

It is not easy to make sweeping generalizations about why none of the research efforts described in this chapter ever made much headway. There were many specific problems such as unwillingness to spend the required amount of money, concern over who would own the technology that might be produced, uncertainty over federal oil shale policies. Doubt of the economic feasibility of the various schemes, though, was always a secondary consideration, and often it was not a factor at all.

Implicit throughout the chapter, however, is an awareness by oil companies that oil shale development could potentially imperil established oil industry operating procedures. Sometimes this awareness has manifested itself in efforts simply to subdue research, especially by the government; sometimes by indifference toward and unwillingness to participate in proposed research ventures, especially where adequate provisions to safeguard the oil industry's interests did not exist; and sometimes, where such safeguards did exist, by modest explorations whose aim is insurance protection against the possibility that somebody else might cause an oil shale industry to spring up or that the government might devise a development program.

The research project described in the following chapter is fundamentally different from the others. With the exception of the abandoned Union Oil program, it represents the only instance where major oil companies have shown apparently substantive interest in shale development. It is too soon to make final conclusions, but recent events in this project may be a harbinger of a basic change in the oil industry's attitude toward oil shale development, from resistance and indifference to acceptance of its inevitability. If development must come, as a few of the more far-sighted oil men (still a decided minority) appear to feel, then the wisest course for the industry is to make sure it is in control.

Chapter 4:
The only presently active research effort
was started by a small independent company.
Then it was racked by personality clashes
and power struggles.
Now it has been taken over
by a large oil company.

SENATOR HART: The obvious question is this: You three people have put in $47 million notwithstanding all the uncertainty about what the Government is going to do. Now, are you foolish, or should I put some salt on the message we have been hearing for a week?

MR. WINSTON: I am not sure I understand, sir.

SENATOR HART: Well, who is right? Are you foolish? Is it folly for you to dump $40 million to the winds because the Government policy is vague or are these people who tell us: "We cannot put any money into this until the Government decides what it is going to do with the oil shale" rationalizing their lack of activity?

MR. WINSTON: May I try to answer that? In our own estimation, we are not foolish.

Senator Philip A. Hart of Michigan and other members of his Senate Subcommittee on Antitrust and Monopoly had spent several days during the committee's oil shale hearings in 1967 listening to a large number of witnesses contend that for a large corporation to put a lot of money into oil shale research and development would be foolhardy. Not only was oil shale production of marginal economic attractiveness, they had said, but the government was blocking all possible progress by refusing to devise a development program. Not only that, the depletion question and the import controls and quotas questions were unsettled. And there wasn't enough private shale land available. And so forth. Then there had appeared a short, dapper, young lawyer named Morton M. Winston, executive vice-president of The Oil Shale Corporation, a small, publicly owned firm which calls itself "Tosco," who had testified that his company and its associates had not only spent $47 million in trying to come up with a workable, commercial shale oil production process, but had spent 12 years in the effort, and furthermore they believed the shale oil business was a very profitable one to be in and they had plans to put into operation within three years the first commercial shale oil plant in the United States in modern times, which would produce 58,000 barrels of shale oil daily.

Winston never did respond to Hart's insistent questions about "Why does it make sense for you to put in $40 million to develop, when it does not for them?" other than by remarking that "It is an exceedingly hard question for me to answer." Tosco people, though, are all too aware that their refusal to capitulate to the allegedly insurmountable obstacles to oil shale development has caused them to be considered mavericks, and has subjected them inevitably to the scoffing and derision of the more conventionally-minded. The scoffers do have some justification for their opinions. For despite a continual display of optimism and several specific predictions of specific dates, Tosco has yet to build its long-awaited commercial shale oil plant.

The Oil Shale Corporation was the creation of Herbert E. Linden, a civil engineer living in Beverly Hills, California, who had been born in Stockholm and emigrated to America at the age of 18. Linden, a large, burly man—he was well over six feet and weighed 250 pounds—displayed many of the familiar traits of the Progenitor/Promoter of the Great Idea: ebullience, optimism, and an ability to convince listeners to disregard immediate problems in favor of the future's certain bountifulness. In the course of a variety of endeavors, Linden made frequent trips back to Sweden where he became acquainted with Olaf Aspegren, who had devised a method of producing shale oil that was radically different from anything ever tried in the United States, or anywhere else for that matter. Though Aspegren had been unable to sell his idea which he called the "Aspeco" process to the operators of Sweden's oil shale industry, Linden realized that it offered several potential attractions over the conventional method of moving hot gases through beds of crushed shale. Its key was the use of "thermospheres," small preheated balls, which were later to cause the concept to be called the "hot balls" process. Aspegren proposed to mix the hot balls with crushed shale in an enclosed chamber. Heat from the balls would retort the shale, and the oil vapors would be drawn off and condensed in another chamber while the cooled balls were reheated by coke from the residue of the retorting. Since everything took place in enclosed containers, the considerable loss of valuable hydrocarbons which occurs when shale is retorted by burning could be avoided. Undesirable chemical reactions with the air which clog up conventional retorts and cause the condensed shale oil to be very viscous could be eliminated. The method would work best on rich shale when other retorts work well only with leaner shale. And while other methods require as long as an hour between the time the rock is introduced and the time it is retorted, the hot balls process, by using much higher temperatures, solid-to-solid heat transfer as well as the enclosed chamber, could be much quicker and much more efficient.

Linden obtained North American rights to the idea in 1952, but his initial efforts to interest friends, despite aggressive proselytizing with a table-model retort, were unavailing until the summer of 1954 when a mutual acquaintance, a cousin of John Jacob Astor, introduced him to Huntington Hartford, grandson of Great Atlantic & Pacific Tea Company founder George Huntington Hartford, who was vacationing in California. Hartford, whose fondness for new and unusual ways to invest his considerable inheritance is equaled only by the inability of most of them to yield him any significant return, was immediately attracted. He contacted Rulon K. Neilson, a longtime friend who is president of Skyline Oil Company, a small Salt Lake City-based producer of oil, gas, and other minerals, and asked him to check the idea out. After several months of investigation, says Neilson, "I told Hartford I think we ought to have a go at it." But Neilson's efforts to raise additional capital from his friends in the oil business were no more successful than Linden's fund raising. "They all told me oil shale was in another generation, if not another century," he says. The only other backer to come forth was Henry Ittleson, Jr., currently chairman of C.I.T. Financial Corporation, which his father had organized. (Ittleson later sold his stock.) In 1955, with total capital resources of about $265,000, The Oil Shale Corporation was officially founded.

Linden's first move was to hire the Denver Research Institute to set up a $250,000 experimental program. The DRI tried to interest several oil companies in contributing to the project without success, but by 1957 a small pilot plant capable of processing 300 pounds of oil shale/hour had been put into operation, and it was then scaled up to a 2,000 pound/hour model. The Aspeco process survived the testing with one significant change: Aspegren had envisioned a counterflow idea whereby the hot balls and the shale would move in opposite directions. This proved impractical and a new system was devised where both would move together in a horizontal rotating drum. Though Tosco by this time had acquired all of Aspegren's patents, a new set had to be applied for. In 1958, a DRI official estimated that the process, now called the Tosco II process, could get a barrel of shale oil to the West Coast for between $1.42 and $1.98.

The first of Tosco's several periods of trauma commenced not long afterward when Herbert Linden, then in his early 70s, stumbled in his bathroom and hit his head against a cabinet, opening up a severe scalp wound. A devout Christian Scientist, he refused medication. The wound became infected, and he became seriously ill. This provoked much concern among members of Tosco's board of directors, which had been joined by Alan M. Strook and David Sher of Strook & Strook & Lavan, Huntington Hartford's law firm, and Nathan W. Levin, a Wall Street investment counselor and investment advisor for Lessing J. Rosenwald. Rosenwald, whose family has long held a major interest and occupied executive positions in Sears, Roebuck and Company, had become a major Tosco stockholder on the advice of David Sher. Except for Neilson, no one on the board had had any direct experience in the oil business. After much discussion, the board members decided they would have to remove Linden as chief executive officer and recruit a successor. Linden's departure, in the view of some Tosco stockholders, would have had to come eventually anyway, for while it was generally agreed that the man was a brilliant engineer, conceptualizer, and promoter, it had become clear that he was somewhat lacking in the more practical aspects of operating a business. The Tosco process was beginning to show signs it might never emerge from the laboratory.

Rulon Neilson was selected to break the news to Linden. "He took it very well," Neilson recalls. "I think he sensed what we were going to have to do." Neilson and others embarked upon a search for a replacement and came up with Hein L. Koolsbergen, a native of Holland, graduate of the Royal Dutch Merchant Naval College, former executive with Royal Dutch-Shell Group, Cities Service, and Southern Natural Gas Company, who at the time was coordinator of oil operations for Newmont Mining Corporation, which had a sizable petroleum division. He had just finished writing a report recommending that Newmont get into the oil shale business. (It didn't.) Koolsbergen was appointed president and chief executive officer of Tosco in the spring of 1961, and Linden was made chairman of the executive committee. Linden's health continued to deteriorate, and though he eventually submitted to massive doses of antibiotics, he died that December. It had always been Linden's idea that Tosco would derive its revenues from licensing its technology to other oil shale developers and it was to this end that $2 million had been spent on research through 1961. Koolsbergen decided that Tosco should prepare itself to be a developer too and set out to acquire oil shale land and water rights. By 1967, $10 million had been spent on 11,141 acres of fee shale land. Koolsbergen also brought in a new management group and commissioned a broad series of studies by outside consultants on mining (DeWitt Smith & Company), refining (Union Oil of California), and other areas, all of which were to be coordinated by the Ralph M. Parsons Company, a large Los Angeles engineering and construction firm. Parsons then prepared a feasibility study for a commercial complex which showed, said Tosco, that "commercial shale crude oil production on appropriate oil shale reserves was not only commercially feasible but, as an economic matter, an attractive investment." Koolsbergen felt, though, that Tosco would be unable to take such a step all by itself and he began a canvass of some of the major oil companies to see who might be interested. He soon elicited the interest of Standard Oil of Ohio, one of the most crude-deficient corporations in the industry which produced under 30 percent of its refinery runs. Standard of Ohio suggested that the two companies might beneficially be joined by someone with mining experience, and after more canvassing they signed up The Cleveland Cliffs Iron Company, a diversified minerals producer whose main business was mining and pelletizing iron ore.

After lengthy negotiations, there was formed between July and September, 1964, Colony Development Company, a joint venture in which Sohio Petroleum Company, a subsidiary of Standard of Ohio, held a 40 percent interest, and in which Cliffs and Tosco held 30 percent interests. Sohio purchased 500,000 shares of Tosco stock—about 10 percent of the total outstanding and reserved for conversion of debentures—plus options on another 600,000 shares, and Edward F. Morrill, a senior vice-president with Standard of Ohio in charge of new product development, was made president of Colony. Colony's objective, the venturers said, was "to obtain specific information and data to enable its owners to make a decision relative to the economic feasibility of constructing and operating a commercial shale oil facility using the best process that can be developed or otherwise obtained." Specifically, they intended to build a "semi-works" or prototype plant using the Tosco process with a "throughput" of 1,000 tons of rock per day to test the technology. No one was committed to proceed with a commercial plant unless it approved intermediary steps. Tosco, meanwhile, remained exclusive licenser of all technologies. In a report to the Oil Shale Advisory Board in November, 1964, Tosco said it expected eventually to be able to produce shale oil for $1.25 to $1.30 a barrel, including depreciation and hydrogenation to remove impurities. Capital costs would run between $1,600 and $2,000 per barrel of daily capacity. Finally the company made its first prediction of the date of commercial operations: It was Tosco's "present expectation" that a 50,000 barrel/day plant would be "in production in 1967."

Construction of the semi-works plant began in Parachute Creek canyon on an 8,300-acre plot of land just a couple of miles north of Union Oil's shale facility that the joint venturers had obtained on option from Dow Chemical Corporation, which had originally purchased it as a future reserve of raw materials for its petrochemical business. By the spring of 1965, a looming, 190-foot-high red retort had been erected and attached to a long conveyor belt-like apparatus that led up to a cliff-face mine, and the plant had begun producing shale oil. Several technical difficulties were encountered. The ⅜-inch-thick ceramic balls began cracking and breaking under the strain of the increased capacity and several months were spent in a complete redesign by Coors Porcelain Company, a subsidiary of the Coors Brewery. Much of the shale oil turned out to contain a sediment which had to be eliminated. The retort, though, operated at a high efficiency, yielding 96 percent of the shale's available hydrocarbons against a yield from other processes of around 75 percent or less. While the shale oil from gas combustion-type retorts is viscous, with a "pour point" of about 90 degrees F., the oil from the Tosco retort flowed well even at 32 degrees. Cleveland Cliffs achieved good results with the mine, a "room-and-pillar" arrangement similar to that at Anvil Points but with a large improvement in costs. Colony at the same time acquired a total of 36,657 acres of shale land, plus water rights, enough to sustain 250,000 barrels of daily production for more than 20 years.

Beneath these rosy-fingered achievements, though, there was trouble, stemming from a growing difference of opinion between Tosco and Sohio personnel. Sohio was officially in charge, but some Tosco workers felt that the pace being set by Sohio was something less than aggressive and expeditious. Specifically, they resented the fact that Colony President Edward F. Morrill of Sohio had established his office in Denver,

200 miles away. Whenever problems developed, operations had to be suspended while somebody tried to reach Morrill on the phone. Tosco engineers lived right on the site in trailers.

Tosco, it must be understood, was, and is, in a very special financial situation. Outside of some interest it earns from investment of current assets, the company generates no income whatsoever, and survives only by means of periodic contributions from its backers, in the form of purchases of additional stock and notes. More than ten years after the company's founding and $12.3 million worth of research expenditures, shale oil production was only a dribble. In the meantime, some rather high-powered financial interests had become interested in Tosco. Southern Natural Gas, a large gas transmission company for which Koolsbergen had worked, had considered providing long-term financing in return for 51 percent of the stock but had backed out. Lehman Brothers, the large Wall Street investment house, though, had made a big stock purchase and had placed two representatives, James W. Glanville and Edward L. Kennedy, on the board. Lehman in turn had successfully solicited an investment from Aquitaine Oil Corporation and Auxirap Corporation of America, subsidiaries, respectively, of Société Nationale des Pétroles d'Aquitaine and Entreprise de Recherches et d'Activités Pétrolièrs, two large oil companies controlled by the French government. A sizable portion of the outstanding stock is publicly traded on the over-the-counter market. "Tosco's problem is that they can't afford to stand still for a minute," says an executive with Colony. "They've always got to keep moving."

The simmering tension between the two Colony partners escalated into a series of harsh encounters between Morrill and Koolsbergen, both strong-willed and tough men. At one point, Koolsbergen angrily suggested to Charles E. Spahr, Standard of Ohio's president, that Morrill be ousted from Colony for incompetence, a request not likely to produce harmony since Morrill was a close personal friend of Spahr's. Members of Tosco's board, especially the men from Lehman Brothers, regularly castigated Standard of Ohio's executives at board meetings. One member referred to Spahr as a "psychopath." Relations among the men up in Parachute Creek were not much better.

Despite these troubles, the joint venturers did manage to agree during the spring of 1966 that the time had come to prepare feasibility studies for a possible commercial plant. Koolsbergen said that Ralph M. Parsons should make the study. Morrill maintained it would be worthwhile to bring in someone less closely tied to the current project who might be more objective. Koolsbergen acquiesced, and Morrill chose Bechtel

Corporation, a large engineering firm which had participated in the erection of the semi-works plant. To everyone's shock, the capital costs for a commercial plant estimated by the Bechtel study were much higher than previous figures, so high that the plant would be only marginally profitable. Tosco engineers blasted the report and claimed it was full of deficiencies, that its analysis was based on Sun Oil's high cost overruns on its tar sands project which had no relevance to Tosco's process and oil shale. The Sohio engineers examined it carefully and said they found it sound. Tosco executives said that even if there was a capital costs problem, the way to solve it would be to keep running the retort while making further studies. Sohio people, supported by Cleveland Cliffs, argued that Colony should shut down the retort, go back to the laboratory and make further studies with small-scale bench models. Tosco said this might delay the project a year. Sohio said if Tosco didn't like it, maybe they should run the retort all by themselves. After a lengthy meeting in New York with Tosco officials, Morrill tersely announced that Sohio was pulling out. In September, the joint venturers announced that Colony was being placed in an inactive status and that Tosco was assuming sole responsibility and financial obligations for the semi-works plant. Sohio and Cliffs said they would continue to run and finance the mine.

The schism was predictably clothed with corporate euphemisms by the various parties. Standard of Ohio's Annual Report blamed it on "changes in tax laws including suspension of the investment tax credit, the high level of activity in plant construction which created long delivery schedules, and the tight money market conditions." Morton M. Winston of Tosco was somewhat more candid before the Senate Antitrust Subcommittee in April, 1967. The problem he said, was "severe difficulties associated with the problems of joint management of a large developmental project."

Others perceived an assortment of scurrilous motives on Sohio's part. Some large Tosco stockholders and others felt that Standard of Ohio was pressured by other major oil companies who were anxious to hold back progress on oil shale research or pick up Tosco's shale land at a fire sale. In the letter that Dr. Tell Ertl wrote on December 3, 1964, to the Oil Shale Advisory Board about Union Oil's experiences with Gulf, he stated that:

My understanding is that similar types of proposals [like that made by Gulf to sell Union cheap foreign oil] also have been made to executives of the Colony Development Company associates. So far the terms have not been good enough. Some of those executives have told me that in their analyses of the world oil situation, importing groups no longer can guarantee sufficient cheap oil, at low enough prices, for long enough periods to compete with Colorado oil shale. We shall see.

Others agreed with Herbert Linden, Jr., son of Tosco's founder and a large Tosco stockholder, who told me, "Sohio really tried to scuttle us, those bastards. They wanted to take over the whole thing themselves."

Whatever the complete reasons, Tosco did come very close to being scuttled, for the firm after Sohio's withdrawal was without funds to continue operations; $531,000 had to be borrowed from some of the executives and directors to meet payrolls and payments on the various plots of shale land on which Tosco held options. Tosco's stock skidded downwards, a slide assisted by a decision by Standard of Ohio to sell out most of its large block. With the aid of Lehman Brothers, Tosco in November obtained a $3.5 million, six-month loan from Aquitaine Oil Corporation, and picked up another $750,000 from the First National City Bank, which had often helped the company out with short-term funds. The money bind was not over, though, for both loans came due the following spring. To raise new funds and repay the loans, Tosco prepared a new stock issue to be offered to existing stockholders. Due to their initial investment agreement with Tosco, the two French companies who own about 12 percent of the outstanding stock have the right to purchase an amount of any new placement on a pro rata basis, and also to buy proportionately any shares in such a placement not subscribed to by others. Consequently, said the prospectus for the offering, if insufficient present shareholders decided to buy the new stock, the French companies conceivably might be able to pick up enough shares "to enable them to elect the entire board of directors of the company." It was not a prospect for which Tosco's other backers had much yearning, and fortunately, enough new money was forthcoming to buy nearly 8 million new shares for $1 apiece, though the offer announced March 8 to expire on March 31 had to be extended to April 7.

Meanwhile, back at Parachute Creek, Tosco had started up the retort (and produced 66,000 barrels during 1967, Tosco engineers point out, against 10,000 produced during Sohio's stewardship) and commissioned further cost studies from Ralph M. Parsons and from Stearns-Roger Corporation, a Denver firm which had built Union Oil's plant.

In his testimony before the Senate Antitrust Subcommittee on April, 1967, Morton Winston said the new "definitive estimate" indicated a commercial plant producing 58,000 barrels/day would cost under $130 million, including hydrogenation treatment and by-product recovery facilities. In comparing this to estimates made prior to construction of the semi-works plant, Winston stated: "These costs are slightly above those originally projected, principally on account of increases in equipment and related costs since 1964. However, the increases are offset by increased capacity over that originally estimated, as well as by substantially increased by-product recoveries and values. In short, the criteria established for the project in 1963 have been met." Then he added: "Tosco's program calls for the first 58,000 daily barrels of production to be onstream in 1970, and we believe that this objective is reasonable and attainable."

Tosco, per the schism agreement, sent Sohio the results of the new study, perhaps harboring the hope that Sohio might be enticed into rejoining, but since Tosco declined to divulge any of the backup data for the study, say Sohio officials, they decided to stick by the Bechtel estimates. When asked at the Bureau of Land Management hearings in 1967 whether he was aware of Tosco's 1970 prediction, Edward Morrill testified he had learned about it only through press releases and added that "We are not involved in any agreement to share or be a part of a commercial plant."

In spite of Winston's brave front, and in spite of a statement in Tosco's 1966 Annual Report that "The year 1966 was an eventful and productive one, in which we achieved some of our most fundamental objectives," the Colony breakup was a serious public relations blow to Tosco and its already tenuous credibility. But there was more adversity ahead. The problem was once again a clash of personalities, this time between the Tosco management brought in by Koolsbergen and Huntington Hartford, who had been serving as Tosco's chairman of the board since the late 1950s. Though apparently capable of some emotionalism, such as in their private dealings with Sohio, Tosco's top managers are generally taciturn, cautious, circumspect, and rather humorless. They are aware that continued support from Wall Street, as well as avoidance of criticism from the Securities and Exchange Commission, which carefully scrutinizes public predictions by speculative companies, require that Tosco's public stance be a model of discretion. The company frequently takes pains to point out, for instance, that it is highly unlikely that shale oil production will ever equal more than a small segment of the crude oil market and that fears that shale oil will "inundate" the market are "baseless." (Another probable reason for this stance is that Tosco believes the future licensees of its technology will be oil companies.)

Huntington Hartford, though, is a celebrity, whose social activities, rather kaleidoscopic marital life, and views on the world in general are closely followed by the press. The failure of his business endeavors, such as *Show* magazine (recently reincarnated and then reburied) and the Gallery of Modern Art, has subjected him to some public derision, and he is considered by many to be the predictable product of a wealthy family who idly sets about squandering his inheritance, a "Horatio Alger in reverse," as he himself once admitted. "I have tried to use my millions creatively," he once wrote in *Show*. "The golden bird, coming to life, has sometimes wriggled out of my hand and flown away."

Hartford's position as chairman was never viewed with rapture by Koolsbergen's new management team, but they tolerated it, if for no other reason than he had, after all, been one of the founders of the company, had occasionally made personal loans to enable it to meet payrolls, had voted dutifully for the ample salaries and stock options desired by Koolsbergen and the other new men, and, with over 900,000 shares, was the company's largest single stockholder. Hartford understood their feelings about him, and tried to remain apart from the daily management of the company as much as possible. But as the rift with Sohio grew larger, he became increasingly frustrated, and he started to intercede by discussing the matter with Charles Spahr and trying to arrange a *rapprochement*. He told the Tosco management the feud was very damaging to the company. Koolsbergen, upset over the intervention, told Hartford to stay out of it, that they could handle things by themselves. Koolsbergen was also disturbed by occasional off-the-cuff statements from Hartford on Tosco and oil shale that would appear in the press without warning, for Hartford is always good copy to a reporter, and when interviewed he is generally casual and frank. "The board has imposed certain limitations on the conduct of its officers," says a Tosco executive. "We have to be agonizingly careful in public disclosures for every disclosure can weigh in the minds of unsophisticated investors. However everything Hartford says, even if he mumbles it in the bathroom, gets reported and it reflects on our corporate conduct."

The growing disenchantment at Tosco with Hartford coincided with the rising influence of Lehman Brothers, principally Senior Partner Edward L. Kennedy, and the rising level of aspirations of Nathan W. Levin, Tosco's vice-chairman of the board and treasurer, who after spending many years working for someone else's money, was growing anxious to stake out his own claim. Levin made it quite clear to Koolsbergen and some of the board members that he would not refuse if they should happen to elect him chairman. Accordingly in 1967, Koolsbergen asked Hartford to resign. Hartford was deeply hurt. He was additionally somewhat shocked at the apparent approval of Koolsbergen's request by Lehman Brothers, through whom he had made his several large sales of A & P stock. But he refused to leave. Relations became icy, and the temperature dropped ever further when Koolsbergen learned Hartford was continuing to "meddle," as one Tosco executive called it. By early 1968, after a year of running the retort on its own and making substantial engineering modifications, Tosco had shut it down again and had begun a series of very difficult and complex negotiations with Atlantic Richfield Company (Arco), the eighth largest United States oil company, which despite its holdings on the North Slope had rather suddenly become extremely interested in oil shale. Tosco hoped that Atlantic Richfield would become its new partner. Then, one day, Hartford happened to have a brief talk on the phone with Robert O. Anderson, Atlantic Richfield's chairman. Koolsbergen learned of the call, and he and a couple of other Tosco executives immediately flew down to Paradise Island, Hartford's island resort in the Bahamas (which he later sold) and severely reprimanded him. Koolsbergen informed other board members of the incident, and in July, 1968, the board voted to remove Hartford as chairman and replace him with Nathan Levin. Edward L. Kennedy was elected vice-chairman. (No mention of the impending rather major change in the company's organization had been made at Tosco's annual meeting which had been held just a couple of weeks previously.) After the vote, Hartford immediately submitted his resignation because, he says, "I didn't want to be in the position of being fired."

He then issued an angry public statement denouncing Tosco's management for acting in a "high-handed and arbitrary manner" in its dealing with Sohio, and charged Edward L. Kennedy with making "derogatory statements at board meetings about Charles Spahr with whom he had conducted a 'personal vendetta.' " At a press conference, he accused Tosco's top officers of "mismanagement" and said he had retained Theodore Sorensen, former advisor to John F. Kennedy and now a partner with Paul, Weiss, Goldberg, Rifkind, Wharton & Garrison, to discuss the possibility of a law suit. He claimed that Koolsbergen, Kennedy, and other Tosco officers had recently sold large holdings of Tosco stock. Finally, he said that Tosco's feud with Sohio was impeding negotiations with Atlantic Richfield (some of the top officers of Sohio and Atlantic Richfield are friends).

All of this airing of soiled linen was greeted by Tosco executives with the most extreme horror, for not since the company's founding had its public image and reputation sunk to such a low point. They realized, though, that continued enmity between them and Hartford would serve no useful corporate purpose and could do a great deal of harm. They commenced peace overtures, and a few months later the squabble was sufficiently subdued for Hartford to consent to be named "Co-Chairman" of the board. He is listed in the 1968 Annual Report below Levin, Kennedy, Koolsbergen, and James W. Glanville.

Negotiations with Atlantic Richfield meanwhile continued to progress, and in a very complex series of agreements announced at the end of 1968 and during early 1969, Arco contracted to pay $24 million— $8 million each to Tosco, Sohio, and Cleveland Cliffs—for a 30 percent interest in "Colony Development Operation," successor to Colony Development Corporation, leaving Arco and Sohio with a 30 percent share and Tosco and Cliffs with 20 percent. Arco's participation gave them a 30 percent portion of Colony's 32,992 acres of shale lands, water rights, equipment, plus rights to use technology. Arco committed itself to an 18-month research program, which would cost all members of the venture a total of between $12.5 million and $17 million—this amounts to a maximum of 14 percent of Arco's 1968 R & D expenditures—and Arco assumed the role of "operator" of the Colony plant, meaning it had complete operational responsibility. Tosco remained exclusive licenser of all technology. Including a $27 million purchase of an 80 percent interest in 10,000 acres and a 50 percent interest in another 1,000 acres from Equity Oil Company in February, 1968,* Arco, which had never before invested in oil shale, was thus obligating itself to payments which could run as high as $56 million, though the expenditures were spread over periods as long as 25 years.

---

* A group of dissident Equity Oil stockholders led by Henry H. Patton, a New York investment banker, waged an unsuccessful proxy contest in May, 1969, charging that, among other things, the price received by Equity was inadequate and that Arco is under no obligation to do anything with Equity's land except sit on it. According to the terms of the deal, Arco is obligated to pay about $1 million a year for the first ten years and the balance, subject to certain conditions, over another 15-year period. The dissidents are currently suing on the grounds that the deal was not cleared with the company's stockholders.

Such an apparently ambitious step toward oil shale development, dwarfing the aggregate efforts of the rest of the oil industry, by none other than the co-discoverer of the huge North Slope oil deposits, has understandably provoked the greatest interest and debate among those persons with an emotional and financial interest in oil shale. Some are very skeptical and contend Arco is just out to sabotage Tosco. Most point out that Atlantic Richfield, under the leadership of Chairman Robert O. Anderson, is generally considered one of the very few imaginative, innovative, and far-sighted major oil companies—before Humble and Arco made their Alaskan discovery, most companies had written off the area as all but worthless. In their opinion, Arco believes that, despite the North Slope, shale oil will inevitably enter the market sooner or later and that a substantial position in sizable tracts of well-situated shale land and the only reasonably well-developed technology is sound corporate strategy. A few people feel that since Atlantic Richfield has taken the first major step, other oil companies will come to realize that despite the lack of a federal policy it is in their interest to take charge of the situation and ensure that the shale is developed in their own best interests. The consensus of the large oil companies who were surveyed for this book and questioned about the development was that, at least immediately, they had no plans to alter their wait-and-see stance. "Atlantic Richfield is a mystery," said a Mobil Oil executive. "I can't help but feel that this is just a case of a powerful chief executive putting pressure behind a pet project."

Tosco executives, for their part, are overjoyed at the development, which has resuscitated their tattered image and elevated their balance sheet to as near a state of health as it has ever experienced. They are at work using the new cash to expand into other energy and raw materials fields to provide Tosco for the first time with a protective cash flow, and they are making investigations of oil shale deposits in Australia. In early 1970, Tosco bought a small oil refinery from Signal Oil & Gas. Though the bulk of Atlantic Richfield's committed spending so far has been on land acquisition, Tosco people feel that Arco has every intention of proceeding ahead expeditiously. But just in case, there are several safeguards written into the contract. Unless firm decisions are made by September, 1970, for instance, on proceeding with a commercial plant, Arco's interest in shale land and Tosco's technology can begin to decrease, and unless a commercial plant is operating by April 1, 1973, at the latest, Arco's fees for using Tosco technology may increase at Tosco's option. The expectation among some at Tosco is that, unless technological problems develop, a commercial shale plant should be running by 1973, allowing for a two-year construction period.

Talking with Atlantic Richfield, one does not sense the same feeling of urgency that apparently exists at Tosco. "We're not dragging our feet," explains Louis F. Davis, executive vice-president in charge of production, "but we're not making haste either. We're going to bring it along in an orderly manner." Arco's reason for being in shale, Davis says, stems from its detailed forecasts of supply and demand over the next 20 years, which show that "we're rapidly running out of oil in this country. Consumption is going up and up, and we figure that sometime near the end of the 1970s, we're going to need synthetic crude from oil shale, tar sands, and eventually coal because there just will not be enough of the liquid variety." (Arco, besides work on shale and tar sands, is investigating coal and moving into nuclear energy.) The North Slope may contain more than ten billion barrels, he goes on, but "there is only enough there to supply us for two or three years." It is conceivable that the gap between consumption and domestic supply could be closed by increasing imports but, he contends that "this will create an unfavorable balance of trade and a big national security problem. Let's face it, you depend too much on imports and you get at the mercy of people who hate your guts. We're saying it's a smart thing for the country to maintain the import quotas and we think the government will agree with us." Davis admits, though, that "our foreign crude holdings are minuscule compared to some of the others, such as Humble."

While Davis believes that eventual production from shale deposits could be "several million barrels a day," he does not think this will have an appreciable effect on the price of crude oil. For one thing, he thinks Tosco's cost estimates as "much, much too low. It's going to require $3 oil to make a shale plant go." Hiram Bond, who is running the Colony plant for Arco, concurs, saying that Tosco's figures are "blue-skyish." Davis also anticipates that, as in other areas of the oil business, production from shale will be geared to the existing crude oil supply and demand situation. If domestic crude reserves deplete faster than anticipated, then Arco would speed up development efforts, he asserts. If new sources of crude are discovered, then they may proceed less rapidly. "You have to keep flexible," he says. But if everything goes along according to "our schedule," there will be substantial production from shale by the late 1970s—in an August, 1968, interview in *Time,* Robert O. Anderson declared that "the first generation [oil shale] plants are five to seven years off. Large scale production is ten to twelve years away."—and Davis is unconcerned therefore about the delay imposed by the Alberta government on Syncrude's tar sands plant startup. Once substantial shale oil production begins, Davis anticipates initiation of a federal leasing policy based on open competitive bidding, probably at regular intervals just like that for the offshore lands. The main bidders, he figures, "will be the people who now own the land, the oil companies." He disagrees with many of the reasons offered by other oil companies for staying away from shale, such as the lack of a government policy or the threat of government competition. The only reason he can offer for Arco's interest in the face of general apathy among the other oil companies is that "I feel we have a little more vision than the rest of the industry. I'm telling you," he concludes, "that place'll be a beehive of activity one of these days." The bees, however, will all be from the same hive.

As for Sohio, a new source of crude oil may now seem somewhat less urgent than when the company joined Colony. In June, 1969, after it had agreed to remain in Colony under Atlantic Richfield's leadership and bear a share of the research cost, Standard of Ohio announced agreement on terms for a merger with British Petroleum Company, the world's third largest crude producer which holds about one-fifth of the world's total petroleum reserves. As the *Oil & Gas Journal* commented: "The conventional view in industry and financial circles is that BP-Sohio is a corporate wedding made in heaven—it is ideal from all points of view. BP gets the market it needs for crude and an expanded position in the U.S.; Sohio solves its long-term supply problem. . . ." Among other things, Sohio would take over BP's extensive North Slope holdings, estimated at 4.8 billion barrels, in return for giving BP a 25 percent interest in Sohio, a figure that would rise to 54 percent depending on eventual production from the North Slope. The top percentage is based on production of 600,000 barrels/day, a rather impressive crude supply for a company which in 1968 processed only 174,000 barrels/day at its refineries. The deal, though, was challenged by the Justice Department. "Justice likes to preserve independent purchasers of crude as competitive forces and as markets for sales by independent producers," the *Oil & Gas Journal* said. But in November, 1969, after criticism from Europeans that the United States was unfairly blocking European investment in United States business, a deal was worked out by Justice to permit consummation of the merger if Sohio and BP would divest themselves of some competing gas stations—BP had picked up some stations Atlantic Richfield was required to sell before its merger with Sinclair Oil was allowed.

The only alien on the scene at Parachute Creek appears to be Cleveland Cliffs, which, rather surprisingly, seems to see its future role in oil shale as something more than a miner of shale rock. "Before we came into Colony, one of the major coal companies was offered our spot," says Thomas E. McGinty, assistant to the president, "but all they wanted to do was mine for a fee and make a profit out of the venture. We agreed to do the mining at cost and share the full risk. That's because we're engaged in a long-range diversification program, and the energy business is one we like very much." A logical next step, he said, would be purchase of petroleum refining and marketing companies. Cleveland Cliffs' total assets are only $197 million, about $\frac{1}{20}$ of Atlantic Richfield's. Nevertheless, contends, McGinty, "We are looking to being a big independent shale oil producer. We want to be a full-range competitor."

PART THREE

"[Oil shale] is the most submerged issue in American domestic politics, involving the greatest scandal in the history of our Republic."

—Paul H. Douglas,
Former U.S. Senator from Illinois,
November 16, 1967

"I live under the shadow of Teapot Dome."

—Stewart L. Udall,
Former Secretary of the Interior,
April 26, 1967

Chapter 5:
The Interior Department,
which one would think ought logically
to be encouraging oil shale development,
has had grounds for apprehension too,
namely the dreaded,
omnipresent Teapot Dome Syndrome.

On November 30, 1921, Edward Doheny, Jr., whose father controlled the Pan-American Petroleum and Transport Company, arrived at the office of Department of the Interior Secretary Albert B. Fall carrying a "little black bag" containing $100,000 in cash. Ostensibly the money was a loan from Edward Doheny, a good friend of Fall's, to enable Fall to purchase a ranch in New Mexico. A number of people were later to feel that the loan and subsequent lease of sections of the Elk Hills Naval Oil Reserve to Pan-American was no coincidence. They were also to feel that receipt of $304,000 in cash and bonds from Harry F. Sinclair, whose Mammoth Oil Company received a similar lease on a Naval Oil Reserve near Caspar, Wyoming (known as the Teapot Dome reserve after a weird nearby sandstone formation) was also no coincidence. The result was perhaps the most notorious scandal in American politics. Though he was never tried on receipt of the Sinclair money, Fall was convicted in 1929 of taking a $100,000 bribe from Doheny. (Doheny and Sinclair were acquitted of playing any role in the episode.) He was fined $100,000, which he was never able to pay, and sent to prison. He died, in the words of one newspaper reporter, "a pathetic, broken man."

Albert Fall's ghost, however, continues to haunt the halls of the Interior Department to this day. It takes the form of what has been termed the "Teapot Dome Syndrome" or the "Giveaway Bugaboo," which has been defined by H. Byron Mock, a former administrator for the Bureau of Land Management, as "the recurrent omnipresent fear of the charge of scandal felt by each federal employee who issues a permit or license or other permission that allows private developers to make a profit from public land." "You have your dome here," former Interior Secretary Stewart L. Udall told the Senate Antitrust Subcommittee in 1967. "Teapot Dome is the dome we live under down at the Interior." A few months after leaving office, Udall reflected on his eight years on the job. "The Interior Secretaryship," he told me, "is the most high-risk, dangerous, perilous job in the Cabinet. Just by signing a piece of paper, you can give away the resources of the nation. I've always said there was only one member of the President's Cabinet to go to jail, and he was a Secretary of the Interior." Then he added, with only the trace of a smile, "I can tell you, I'm goddamned relieved *I* got out of there without going to jail."

Secretaries of the Interior have been willing, of course, to lease to oil companies large sections of federal offshore land, which has returned billions of dollars of royalties, rentals, and bonuses to the Treasury. A few oil and gas leases have even been granted on the public oil shale lands for recovery of conventional oil and gas deposits. But the issuance of leases for oil shale, which might have to include such provisions as special incentives to encourage private industry to undertake an expensive research and development effort, is another matter. In the 1965 report of Udall's Oil Shale Advisory Board, Harvard economist John Kenneth Galbraith wrote that leasing the shale land would be "gravely damaging to the public interest" since:

—The cost of development is unknown.
—The cost of production is unknown.
—The recoverable value of oil in the land offered for lease is imperfectly known.
—Given these unknowns the government would be offering a subsidy of unknown value for development of unknown cost promising a return of unknown amount. This amounts to dispersing public property while wearing multiple blindfolds.

Galbraith concluded he was strongly opposed to "helter-skelter alienation of this great public resource." Stewart Udall told the Senate Interior Committee in 1967 that "the big message [Galbraith] had was— there are too many unknowns, and as long as there are unknowns, you run the risk of giveaway—you run the risk of not knowing what you are doing."

Secretaries of the Interior have, without exception, exhibited an extensive distaste for wrestling with these unknowns. "I think it is quite accurate," Udall told the Senate Antitrust Subcommittee in 1967, "to say that the policy or nonpolicy of Secretaries of the Interior [on oil shale] was to say that they were studying the problem." With the exception of two small leases in Wyoming and one in Utah in the 1920s, which were all later canceled, no public shale land has ever been leased. Federal attempts to encourage oil shale development have been all but nonexistent. Even federal survey of the government's own shale land has been at best rudimentary. "Possible upward revisions of the shale oil resources of the Green River Formation may equal what is already known," said Russell Wayland, acting chief of the Conservation Division of the Geological Survey in 1967. "Larger undiscovered, better grade resources will perhaps be found in the deeply buried parts of the Uinta Basin." As Chapter Seven will relate, it was not until 1965 that the extensive existence in the shale deposits was discovered of an aluminum-bearing mineral called dawsonite which may have large commercial significance; and this discovery was made not by the government but by private individuals who through an administrative fluke had obtained permits to drill core holes on public land.

The Interior Department's apprehension over oil shale has been caused by more than simply the general presence of the Teapot Dome Syndrome. Twice during the past fifty years, small tremors of scandal specifically concerning the oil shale lands have threatened to escalate into a mammoth eruption, with a heavy loss of political lives. In both cases, the circumstances are far less readily assimilable and graphic than Albert Fall's receipt of the little black bag—indeed this is probably one of the chief reasons why neither received very much public attention. In order to comprehend them, it is first necessary to review some of the essentials of the mining laws and the remarkable shift in the philosophy of the use of public land around the turn of the century.

The American system of mining laws in many respects can be said to have had its origin the day during January, 1848, when James W. Marshall and Captain John A. Sutter discovered small grains of gold while panning in a branch of the Sacramento River near Coloma, California, thus setting off one of the greatest mineral rushes in history. While thousands of prospectors swarmed over large areas of public land extracting gold, state and federal legislators frantically debated just what to do about this clearly illegal mass trespass. Many thought the land should be licensed or sold or rented so that the government could receive some compensation. Others argued that the miners should be left alone. The latter viewpoint carried with it an attitude toward public land that had persisted since the nation's beginning. A man who occupied a piece of land, and especially a man who cultivated it, worked it, or built his home upon it, was generally considered to have established a kind of de facto ownership. (Indians, though, were regarded as an inferior cultural group with no such rights, and were freely pushed off land on which they had lived for centuries.) This ownership, further, was thought to constitute a kind of reward, for by occupying and developing it, the man was in a sense giving essentially worthless land a value which contributed to the public good. He was, too, helping to extend the hegemony and influence of the nation of which he was a part. While the land owned by a democratic government is owned in theory by the people from whom it receives its legitimacy, the role of the government was felt to be to hold the land in escrow until title could justly be passed on to the citizens. "Land, the gift of God to man," said Missouri Senator Thomas Hart Benton in 1805, "is also the gift of a government to its citizens."

The United States' first public land laws established a system of auction sales to generate sorely needed revenues, but widespread domination of the auctions by speculators and open defiance of the laws by impoverished settlers who became "squatters" on large areas of public land, caused abandonment of the auctions and a return to the free land concept. It was codified in various settlement laws, especially the famous Homestead Act of 1862. In return for a small registration fee, any settler over 21 who was head of a family could receive 160 acres of public land, the only stipulation being that he must raise crops on the land for five years before receiving a "patent" or final title. While most of the settlement laws specifically reserved mineral rights to the government—an old European practice—most of the valuable Eastern coal, iron, lead, copper, and zinc deposits were, with tacit government approval, illegally appropriated as settlement lands.

With Western legislators vehemently opposing any "tax on labor and enterprise," Washington's official reaction to the gold rush became a policy of laissez faire. President Fillmore in 1851 stated that the gold fields should remain "open to the enterprise of all our citizens, until further experience shall have developed the best policy to be ultimately adopted." In the absence of official guidance, the gold miners meanwhile had set up their own codes based on European and Mexican mining laws. These rules, which were enforced with surprising effectiveness, established priority of possession as constituting a prior right, fixed the size of a valid claim, and required that a reasonable amount of work be performed to hold a claim against subsequent claimants. Indeed, their principal purpose was to protect a man who staked a claim and diligently pursued its development from being displaced by a later prospector who desired the same land.

The heavy costs of the Civil War as well as arguments that hundreds of millions of dollars were being taken from federal lands, often by foreigners from Mexico and Europe, without the slightest additions to the federal Treasury, finally produced irresistible demands for legislation. The result, after much Congressional hassling, was the Mining Act of 1866, which was more or less a codification of local practices and, as it turned out, a near total victory for Western views. It declared that "the mineral lands of the public domain, both surveyed and unsurveyed, are hereby declared to be free and open to exploration and occupation by all citizens of the United States" subject to local rules and customs as long as they did not conflict with federal laws. (State laws generally provided that a claimant to a parcel of land, to make his claim valid, must pound in corner boundary markers in the corners of the land, post a notice that he had located the claim, and register it at the county clerk and recorder's office. Despite several attempts, no federal legislation has ever been passed providing for a more centralized recording system.) Additional provisions specified the claim size and stated that if a claimant adhered to various occupancy and development requirements, he was entitled to receive from the Interior Department, the official custodian of the public lands, a "patent" on his claim.

A mining claim is basically a legalized trespass, an exclusive possessory right granted automatically to whomever stakes it to extract the minerals contained in a section of land. This right is subject to challenge by later claimants on the grounds that the first claimant's claim is invalid under the provisions of the mining law. The government, meanwhile, retains actual title to the land. But in granting a patent to a qualified claimant who has a valid claim, the government relinquishes to him full, perpetual ownership to the land, its surface and its minerals. In practice, few mining claimants took the trouble to apply for patents. The interest of most prospectors was the minerals, and after extracting the minerals, they simply abandoned the land and moved on.

While the 1866 law covered only so-called "lode" claims on veins of minerals, the 1870 Placer Act gave miners the right to stake "placer" claims for minerals scattered on or near the surface of the ground, such as gold granules in river beds, and, according to later interpretation of the law, deposits of oil and oil shale. Patents on placer claims could be obtained for $2.50 an acre, a fee to cover administrative costs; patents on lode claims cost $5.00 an acre.

The 1866 and 1870 laws were combined and revised by the Mining Law of May 10, 1872, whose basic provisions are still in effect. It restricted claiming to "valuable mineral deposits" whose physical "discovery" was required before a claim could be considered valid. The claimant's freedom to enter upon public land and extract minerals was essentially unchanged, and the new terminology only became meaningful if his claim were challenged by another prospector or he should desire to apply for a patent. In granting the patent, it was the government's duty to determine whether in fact the applicant's claim was valid. Just how the government was to interpret the new terminology was to engross legions of lawyers over many generations in convoluted squabbles over, among many other things, the validity of oil shale claims and, in some cases, whether certain interpretations of the mining law constituted a scandalous giveaway. The principal issues have been: Just how valuable is "valuable"? And just what kind of a discovery is a "discovery"?

Other important sections of the 1872 Mining Law provided that:

1) though the size of a claim staked by a single individual was limited to 20 acres for each discovery, an association of eight individuals could locate up to 160 acres for each discovery. No limit was placed on the number of claims any one individual or association could stake;

2) claimants were required to perform each year at least $100 worth of development or "assessment" work on each claim in order to maintain the claim's validity. An applicant for a patent had to show he had performed $500 worth of work on the claim. This provision was not merely designed to protect a claimant against other claimants. Since one of the chief purposes of the Mining Law was to facilitate mineral development, assessment work requirements helped make sure that a prospector could not just sit on his claim for speculative reasons and prevent access by a more development-minded prospector;

3) the Secretary of the Interior was given the basic authority to administer the Mining Law.

The free land philosophy which underlay the Mining Law began to change during the late 1800s when the public first became aware of flagrant abuses of the settlement and mineral laws. Vast areas of timber lands, mineral lands, grazing lands were being illegally appropriated and horribly ravaged, usually by large corporations who were thereby acquiring immense power. These corporations, and in fact the free land idea in general, had been instrumental in causing the settlement and development of the West. But now that the West had already been largely settled and developed, people began to wonder whether the old incentives still applied. They began to question for the first time whether private ownership was always better than public ownership, whether it was always consistent with the public good. And as they considered the federal policy of almost uninhibited land grants for such social ends as construction of roads, canals, and, especially, railroads, reclamation of desert lands, establishment of educational institutions, they began to wonder whether the nation's supply of land and resources was really as limitless as had once been believed.

Thus began the great conservation movement, which first flourished during the administration of Theodore Roosevelt. The new views were well articulated in a 1909 report of Roosevelt's National Conservation Commission:

The resources which have required ages for their accumulation, to the intrinsic value or quantity of which human agency has not contributed, which when exhausted are not reproduced, and for which there are no known substitutes, must serve as a basis for the future no less than the present welfare of the Nation. In the highest sense, therefore, they should be regarded as property held in trust for the use of the race rather than for a single generation, and for the use of them by right of discovery, or by purchase. . . . The right of the present generation to use efficiently of these resources what it actually needs carries with it a sacred obligation not to waste this precious heritage.

To these ends, Roosevelt made a series of controversial executive "withdrawals" of valuable public land containing coal, timber, oil, water, and other resources from further private entry, and committed himself, through the 1902 Reclamation Act, to a positive policy of improving and developing public land. Further withdrawals were made by Presidents Taft and Wilson, especially of oil lands, and the Naval Oil Reserves and the Naval Oil Shale Reserves were first established. This provoked another of the fierce battles between Eastern conservationists and Western developers that have occurred throughout American history and, indeed, are still going on. Typical of the Western point of view was a statement by Thomas Burke, a senatorial candidate from the state of Washington during the 1910 campaign:

The people of today have a right to share in the blessings of nature. There is no intention of the West to rob the future but there is a determined purpose not to let a band of well-meaning sentimentalists rob the present on the plea that it is necessary to hoard Nature's riches for unborn generations.

On February 25, 1920, there was a compromise: The Mineral Leasing Act which ended, at least in part, what one lawyer calls the "finders-keepers" philosophy of the Mining Law. It stated that all federal land containing a number of non-metalliferous minerals such as oil, oil shale, coal, natural gas, phosphate, and sodium could no longer be claimed under the Mining Law but could only be leased by the Interior Department in return for payment of rentals, royalties, and bonuses. The government retained final title to the land and the minerals. Restrictions were put on the amount of land any one party could obtain—in the case of oil shale it was one lease per person, association, or corporation with a maximum of 5,120 acres. Royalties were to be distributed in the proportion of 37½ percent to the state in which the minerals were located, 52½ percent to the western-oriented Reclamation Fund established under the Reclamation Act for the building of dams and other projects, and 10 percent to the federal Treasury. (An exception is Alaska which gets a 90 percent share of the revenues.) The Western developers, therefore, were the chief beneficiaries of the revenues under the Act, while the Eastern conservationists now possessed a mechanism to prevent control of mineral lands by private monopolies, to discourage holding of land for speculative purposes, to allow the government to obtain something closer to a fair share of the returns from mineral development than it had received under the Mining Law, and, in general, to allow the people to share in the mineral resources of which they were the owners.

While the Mineral Leasing Act prohibited further claiming of the oil shale lands, Section 37—the so-called "Saving Clause" inserted specifically by oil shale interests—said that valid oil shale claims existing on the date of the Act and "maintained in compliance with the laws under which initiated, including discovery" continued to be valid, and were entitled to receive patents. It will be recalled that at least four million acres worth of claims had been filed during a period of a few years prior to 1920.*

* Technically, the oil shale claims had been filed under the authority of the 1897 Petroleum Placer Act, which had provided that placer claims could be filed on lands containing petroleum that were "chiefly valuable therefor," another

In 1923, a dispute arose between George L. Summers, who had filed homestead claims on some Colorado shale land—homestead laws formally reserved the oil shale and other minerals to the government—and J. D. Freeman and the Standard Shale Products Company, who had purchased some oil shale placer claims staked a few years earlier on the same land. At issue was the validity of Freeman's claims and, more generally, the interpretation of the "discovery of a valuable mineral" phrase of the Mining Law. Court interpretations of *Castle* v. *Womble* in 1894, had established the so-called "Prudent Man" test of a claim's validity: the mineral discovered had to be of sufficient value that "a person of ordinary prudence would be justified in the further expenditure of his labor and means, with a reasonable prospect of success, in developing a valuable mine." In the section of land under dispute, there was a lean, thin layer of oil shale near the surface. Underneath was an area of sandstone and other sedimentary rock containing very little oil shale. About 500 feet below the surface was a second layer of shale, very thick and very rich. Freeman, the owner of the oil shale claims, said that traces of the upper lean shale layer had been discovered outcropping on the land and that from this evidence, a prudent man could "infer" the existence of valuable shale beds below since both were part of the same large deposit. Summers, the homesteader, argued that the two beds were really completely separate, and that however valuable the lower shale layer might be, the lean outcroppings, the only shale whose presence the claimant could be certain of, were worthless, and no prudent man would consider trying to develop a valuable mine.

After several hearings, beginning with the local Interior Land Office and working their way up the departmental hierarchy, First Assistant Secretary and later Solicitor Edward C. Finney ruled in 1924 in favor of Summers. Freeman appealed, and after more hearings, Finney reversed himself. In September, 1927, Interior Secretary Hubert Work approved a decision by Finney in favor of Freeman. The *Freeman* v. *Summers* ruling laid down two very important discovery assumptions that were to become the focal point of many controversies to come:

1) The oil shale in the Green River Formation is more or less one large deposit of fairly predictable dimensions. ". . . (T)he whole body, therefore," it said, "will be commercially developed and is all valuable, and *discovery of any shale upon the land,* being a part of the integral

phrase which has been the subject of much litigation. Later interpretations of this Act by the Interior Department held that it also applied to oil shale lands.

mass that lies below it, is a sufficient discovery to satisfy the requirements of law." (Emphasis added.)

2) A mineral could still be considered valuable, even if its value was only prospective. "It is not necessary," the opinion stated, "in order to constitute a valid discovery under the general mining laws sufficient to support an application for patent, that the mineral in its present situation can be immediately disposed of at a profit." Even though the ruling was issued during a time of immense crude oil discoveries, it said "there is no possible doubt of [oil shale's] value and of the fact that it constitutes an enormously valuable resource for future use of the American people."

Under these postulates, it could be presumed that anyone requesting a patent on an oil shale claim who could show even a small trace of oil shale from anywhere in the Green River Formation would totally satisfy the discovery requirement of the Mining Law.

The *Freeman* v. *Summers* decision infuriated a man named Ralph S. Kelley, who was in charge of Interior's Land Office in Denver. Kelley was a member of a new breed of Interior employee which had first come into being with the rise of the conservation movement and which was later to turn the Department away from its initially Western, pro-resource development orientation. Members of the new breed saw themselves as *protectors* of the public domain, and they regarded with contempt the old belief that the criterion for disposal of public land was whether the public interest required the government to retain ownership. They felt instead that the burden should be on private parties desiring rights to public land to prove that the public interest was benefited by the land's disposal to them. They regarded such private desires with skepticism and disdain and increasingly they contested them with vigor and even ferocity.

Ralph Kelley had already commenced investigations of the validity of some of the oil shale claims, and in 1925 he received permission to intervene in *Freeman* v. *Summers* on behalf of the government. His sympathies clearly lay with Summers, and after the decision had been made he assembled a large volume of evidence to disprove Freeman's assertions, principally that the oil shale deposit was not one single integrated entity, as Freeman had contended. Kelley demanded Finney reconsider. A series of bitter squabbles between Kelley and Finney followed during which Kelley seriously questioned Finney's integrity. In a letter to Ray Lyman Wilbur, who took over as Interior Secretary in 1928 when Work quit to become chairman of the Republican National

Committee, Finney called Kelley's contentions "the ravings or imaginings of a diseased mind. It is evident that the man is egotistical and suspicious of the honesty and motives of other persons who happen to disagree with his views or conclusions." In 1930, Wilbur relieved Kelley of his duties in Denver and ordered him to come to Washington. Kelley arrived in July and promptly telephoned Walter Lippmann, who was running the editorial page of Herbert Pulitzer's New York *World,* and offered to sell his story of the episode. The *World* agreed, for a fee of $12,000, and on September 28, Kelley submitted a letter of resignation to the Interior Secretary. On October 6, the first installment of Kelley's 14-part series was published. The page one headlines read:

KELLEY'S OWN STORY QUOTES WILBUR'S FILES SHOWING HOW FAVORITES GOT OIL BILLIONS

Politicians with Pull and Some
Complaisant U.S. Officials
Are Accused of Aid in
Circumventing Law

$40,000,000,000 Public
Lands is Prize at Stake

800,000-Acre Domain Dwarfs
Teapot Dome in Value—
One-Sixth Now Gone

Kelley's principal charge was that Finney had reversed himself on *Freeman* v. *Summers* due to pressure from a number of large oil companies, who had acquired considerable acreage in pre-1920 oil shale claims which, if Finney's decision were upheld, would easily qualify for patents. As evidence, Kelley described a public hearing that had been arranged by Colorado's two senators in December, 1926, in Secretary Work's office to discuss the issue of discovery raised by the *Freeman* v. *Summers* case. In attendance were representatives of 13 oil companies, several Colorado congressmen and other politicians, and numerous owners of oil shale claims who had made patent applications. A principal speaker was Freeman's counsel, Robert D. Hawley, who was also attorney for Union Oil of California, a large shale landholder. Neither George L. Summers, nor any of his representatives, was present, for after all the litigation Summers had no money to pay for the trip to Washington. In an effort to tie this in with Teapot Dome, Kelley mentioned that Finney had been Assistant Interior Secretary under Albert Fall and that among those with interests in shale land were the Midwest Refining Company, a subsidiary of Standard Oil of Indiana, which had acquired control of E. L. Doheny's Pan-American Petroleum and Transport Company, and the Prairie Oil & Gas Company, which was being acquired by Harry F. Sinclair.

Kelley made additional charges concerning the issue of assessment work, the $100 worth of labor the Mining Law said had to be expended annually per claim to maintain its validity. Kelley and some other employees at the Denver Land Office had noticed that assessment work was not being performed on most of the oil shale claims. As has been noted, a chief purpose of the provision is to protect a claimant against other claimants. But since the Mineral Leasing Act had precluded further claiming on oil shale land and since the Mining Law had also been designed to ensure that claimants pursue diligently the development of their claims, Kelley figured that the government could become, in effect, a subsequent locator and challenge the validity of the claims on the grounds that assessment work was not being performed. The Mineral Leasing Act after all protected only those claims being "maintained" in accordance with the Mining Law, and the Supreme Court had specifically stated in 1920 that Interior had the power to determine whether a claim was valid or invalid and that if it were invalid "to declare it null and void." Acting on this line of reasoning, the Denver Land Office in the mid-1920s had started a broad campaign of challenges, principally on the grounds that assessment work was not being performed, but also on the grounds of fraud. Many thousands of claims were canceled.

As Kelley described in the *World* series, among the claims voided for nonperformance of assessment work was a 7,000-acre parcel in which Colorado Congressman William R. Eaton, who had done legal work for Midwest Refining, had an interest. Eaton and other claimants made highly vocal complaints to Work, Kelley said, and in 1928 Work told Kelley to halt his campaign, ostensibly to await a Supreme Court ruling on a challenge of Interior's cancellation policy which was working its way up through the courts. Kelley quoted a telegram sent in early 1929 by Colorado Senator Charles W. Waterman to Midwest Refining's Denver attorneys: "In compliance with wishes of yourself and my friends in Colorado interested in shale applications Department of Interior will immediately direct Denver field division to postpone all hearings on pending shale contests. . . ." This holdup, charged Kelley, caused a "virtual paralysis" in his office's functions. In the meantime,

Congressman Eaton was introducing three bills which in effect would have all but granted patents outright to all shale claimants. Though his bills did not pass, between 1928 and 1930 over 70,000 acres of shale land passed into private ownership.

In spite of lengthy detailings of circumstantial evidence, Kelley was never able to show that there had ever been anything as damning as a bribe made in return for the alleged favors. The *World* promoted the series aggressively, but its efforts to syndicate it nationally were, a *World* editor admitted, "almost a complete flop." One editor complained to the *World* that the articles were "too long and too dry and devoid of public interest. . . ." (Having seen more glorious days, the *World* expired in February, 1931.)

Kelley himself was not doing much better. The Justice Department submitted a report to President Hoover which stated that "there is obviously no basis for charging misconduct in making such a decision [*Freeman* v. *Summers*]. The decision seems eminently reasonable, but at most it was merely a matter of judgement on a debatable point." Kelley was publicly denounced by Wilbur, who fired him despite his resignation, and by Hoover, who termed his charges "reckless, baseless, and infamous." Senator Thomas J. Walsh, who had heroically distinguished himself by leading Congressional investigation of the Teapot Dome scandal, eagerly conducted hearings by the Senate Committee on Public Lands and Surveys in early 1931 into Kelley's purported scandal. But he soon found himself bogged down in the labyrinthine complexities of the mining laws. Nowhere to be seen was anything as arresting as a little black bag full of payoff money. After five days of testimony by Finney and Wilbur, Kelley was quizzed briefly about how he had gone about selling his story to the *World* and in the end he was even refused the opportunity to make a statement. The Senate took no further action. Interior during this period had held up oil shale claim patenting. And in April, 1930, Herbert Hoover had "temporarily" withdrawn the oil shale lands "from lease or other disposal," an order which has yet to be rescinded. But after the hearings, the chairman of the Senate Public Lands Committee advised the Secretary of the Interior that "there is no reason why your Department should not proceed to final disposition of the pending applications for patent to oil shale claims in conformity with the law." Patenting resumed, with the principles of *Freeman* v. *Summers* in full control.

Ralph Kelley's luck in the Supreme Court appeared at least temporarily to have been somewhat better than that elsewhere. The assessment work case reached the Court in 1930, and in *Wilbur* v. *Krushnic* the Court ruled that a claimant did not lose his right to a claim for failure to perform assessment work as long as he resumed work before a government challenge. The Interior Department interpreted this to mean that it could challenge claims while assessment work was in default, and in a broad, extremely energetic challenging project, it eventually voided 22,245 claims covering 2,884,091 acres.* While all this activity may have been a reaction in part to the Kelley series, it should be realized that the huge East Texas oil field was discovered in 1930, which virtually dissolved the interest in shale land of most oil companies as well as other claimants. One party whose claims had been canceled did have enough interest to challenge the *Krushnic* decision, though. In 1935, the Supreme Court, in *Ickes* v. *Virginia-Colorado Development Corporation,* revised its stand and ruled that Interior's cancellation activities for nonperformance of assessment work "went beyond the authority conferred by law."

Some claimants whose claims had been canceled requested reversals, and in its 1935 *Shale Oil Company* decision, Interior officially reversed its policies by canceling a Land Commissioner's decision declaring some claims void for lack of assessment work and stated that "other Departmental decisions in conflict with this decision are hereby overruled." In response to letters from claimants, Interior advised its former cancellation decisions were without effect. The stage was set for the beginning of the next alleged scandal.

The interest of the major oil companies in the oil shale lands began after the Second World War when, in view of the predictions of crude shortages as well as the government's research work at Anvil Points, many of them began to feel that shale land might be a wise investment for the future. Agents were dispatched to see what might be available. Buying up shale land was somewhat more difficult than going through the real estate ads in the newspaper. It was a difficult enough problem merely finding who were the legal owners for the old claims, for by this time many of the old prospectors had died or moved away, and though technically a mining claim is a property right that can be passed along as part of an estate, heirs to these interests were often unaware of what they owned. The only records of ownership were musty tract books in

---

* Among those whose claims were voided was J. D. Freeman who had never gotten around to applying for a patent. Summers thereupon reasserted his rights and was granted a homestead patent in 1932.

the county courthouses, and location entries were often confusing and illegible. And even if claims could be acquired, there was the even greater problem of applying for a patent—only a small portion of the land had been patented. Obtaining a patent involves a laborious assemblage of numerous documents and supporting papers and a convoluted trip through the federal bureaucracy. Some of the old oil shalers from the boom period, who had remained in the area despite a rather destitute existence, saw an opportunity at last to make some money from the land that had been a source of frustration for so long. Because of their knowledge of the arcane nuances of the mining laws and their acquaintanceship with the old claimants and their relatives and heirs, they were able to set themselves up as land speculators and agents with whom outside buyers usually found it very convenient to deal. A further advantage to the oil companies of working with these men was that they could thereby disguise their interest in the land, which otherwise inevitably would have driven up prices.

The speculators worked this way: After an arduous study of the county records they would find the names of the original locators—most claims consisted of 160 acres staked by an association of eight individuals—and track down the location of one of those men. The old claimants or their descendants generally regarded the old claims as all but worthless, and were delighted to sell their rights, often for only a few cents an acre. With the one-eighth interest in hand, the speculator was ready for a procedure known as *forfeiture*. First he would perform $100 worth of assessment work on the claim. Then he would advertise in a local newspaper for the other seven claimants to contribute their share. If the other men failed to respond, as was almost always the case, the speculator was free to institute a title suit and obtain a decree granting him sole right to the claim. Even though the *Virginia-Colorado* decision implied that performance of assessment work was not a requirement for maintenance of an oil shale claim's validity, the forfeiture procedure has been upheld by the courts. Between 1952 and 1962, some 94,800 acres of shale land changed hands through forfeiture in Colorado alone.

Some of the speculators dealt in unpatented claims, but most of the buyers desired the security of possessing complete title. After forfeiture, speculators generally would request patents. Though a patent might not be forthcoming for a year or two, the reward was great. Unpatented land during this period was worth only a few dollars an acre, but patented land, obtained for only a $2.50/acre fee to the government, was worth $100 an acre and, during the mid-1950s, the price rose to $500, depending on the quality of the land. Since then, the price has risen even further. The series of deals between Colony Development Company, Dow Chemical, Atlantic Richfield, and Equity Oil during the 1960s involved prices, depending on the exercise of various options, of between $3,200 and $15,000 an acre.

Typical of the speculators was a man named Delos Demont Potter, a large, formidable-looking former college wrestler from Kansas and a graduate of the Northwestern Law School. Swept up in the shale land rush, he formed the Federal Shale Oil Company in 1917. Though the company never produced any shale oil, it did much buying and selling of shale land during the 1920s, and Potter frequently fought Ralph Kelley in an effort to secure patents on the company's claims. During the 1930s, Federal Shale Oil and Potter's large bank account faded away, and he squeezed out a meager existence practicing law in Rocky Ford, Colorado. His enthusiasm for shale did not abate, though, and twice, through lobbying with Western congressmen, he helped defeat bills periodically produced by the Interior Department to require performance of assessment work on the old claims.

After interest in shale land had revived, Potter reactivated Federal Shale Oil Company and assembled a team of mining engineers and others in Denver to buy up and patent some of the old claims. Potter's success in obtaining patents was probably unmatched by anyone. He developed a thorough working knowledge of the Interior bureaucracy, and he would often carry his applications by hand from one desk to the next, both in Denver and Washington. He would sit patiently, waiting for a particular man's initial. If an organizational roadblock developed, he would suddenly shake his fist, and thunderously demand that his rights under the law be granted. Using these methods, Potter's organization, according to *The Rock That Burns,* a book of reminiscences by an old oil shaler named Harry K. Savage,* was able to obtain patents on nearly 100,000 acres, which he then sold to such companies as Union Oil of California, Shell Oil, Sinclair, and Standard of California. With his new wealth, Potter became one of the most flamboyant figures in Denver, dressing in expensive Eastern clothes, driving shiny Cadillacs, and entertaining at the best clubs. He threw annual parties for his oil shaler friends from the West Slope, as the shale country is referred to in

* Savage wrote the book when he was 85. Now, at 91, he is working on a supplement while conducting oil shale experiments in his garage in California.

Denver, to which were also invited local Interior Department employees, a practice which did not injure the smooth if intricate flow of patents. When Potter died in 1954, his estate, despite profligate spending, was said to have been worth well over $400,000.

Another of the old timers is Joseph T. Juhan, who may have been even more financially successful than Potter. Among Juhan's best deals, according to friends, was the sale of a plot to Standard Oil of New Jersey for 50,000 shares of Jersey stock, now worth nearly $3 million. The sale to Jersey was rather ironic, for during the 1930s, he had purchased a block of 20,000 acres from Carter Oil, a Jersey subsidiary, for 25¢ an acre—at the time Jersey probably figured shale land was worthless. Since he had no money to pay Carter, he sold the surface rights to some local ranchers for 50¢ an acre, reserving the shale rights underneath, of course, to himself. Later he sold the block to Standard Oil of California for $300,000. As has been his practice, he retained a ⅛ interest just in case the long-awaited shale industry should spring to life.

Probably the most spectacular oil shale land speculator of modern times is Dr. Tell Ertl, though many of the old oil shalers regard his flashy dealings with scorn and, since he was not around during the 1920s, feel he is decidedly *nouveau*. Ertl, it will be recalled, helped convince Union Oil to build its research plant. He also helped Union acquire shale land, and between 1952 and 1956, Union obtained 10,880 acres through forfeiture proceedings. But during this period, he was picking up a little land for himself. He had long admired two large areas of excellently situated land, one of which was as close to the center of the Piceance Basin as any private land, save for some narrow strips of patented homestead claims along the interior creek beds (and later acquired by oil companies interested in *in situ* experimentation). Between 1954 and 1957, according to Interior records, he obtained through forfeiture at costs as low as 3.9¢ an acre interest in at least 221 unpatented claims covering 35,040 acres. In 1964, he signed an agreement with Shell Oil giving it the option to buy one 21,120-acre block for $42,240,000, or $2,000 an acre, if Ertl were able to obtain patents on the land. Shell agreed to pay all litigation costs, up to $130,000 a year, and pay Ertl (who shares interest in the land with a couple of other individuals) an annual retainer of $50,000 plus a maximum of $100,000 in research fees. Ertl also, according to Interior, sold a plot of 16 unpatented claims to The Oil Shale Corporation for $1,536,000 in 1964 and made a deal on additional acreage for an option for which

Tosco pays him $148,000 annually. While Ertl asserts these deals have not yet made him rich, he did assemble enough money to purchase two years ago a ski resort near Denver called Lake Eldora.

Throughout the 1940s and the 1950s, the granting of patents by Interior on shale land was a relatively routine, if complex, procedure. *Freeman* v. *Summers* and *Virginia-Colorado* were interpreted as guiding precedent. Any claim that contained a small showing of oil shale outcrop was considered to have fulfilled the discovery requirement, and previous cancellation for nonperformance of assessment work were disregarded. But more significant was the spirit of the era, well embodied in H. Byron Mock, who was Regional Administrator in Colorado and Utah for the Bureau of Land Management from 1947 to 1954, and had a prime responsibility for approving applications for oil shale patents. Now a Salt Lake City attorney with numerous mining clients, Mock has long been a Western-style advocate of mineral resource development. Along with other Interior officials at the time, he felt that the somewhat uncertain ownership status of the privately-owned shale lands was a major block to shale development. Shortly after he assumed his post, Interior Secretary Julius A. Krug traveled out to Glenwood Spring, Colorado, journeying part of the way on a Denver and Rio Grande train powered by shale oil. After a meeting with local oil shalers and claim owners he said he would do what he could to remove obstacles to oil shale development that originated in Interior, though he was not ready to take the political risk of putting together a leasing program.

Mock thereafter did everything possible, he related in a January, 1966, *Denver Law Review* article, to "accelerate processing of claims to patent." "Probably most impressive to me," he said, "was the large number of dedicated mining men who had sunk every available dollar into developing and retaining and patenting oil shale claims. Even then sons of those original pioneers were succeeding to the struggle as the original pioneers began to die off." Yet one did not have to be a pioneer, apparently, to gain the sympathy of the Interior Department. Tell Ertl describes his visits to Washington with Union Oil lawyers to help them push along their patent applications this way: "We'd just stop over to pay our respects to the Congressional delegation, and then we'd drop over to the [Interior] Secretary's office to see what was going on and see whose desk our patent applications were on. Often the Secretary would just pick up the phone and tell his people to get a move on. During this period, it helped to have power, and the oil companies could supply that." In 1954, the process was simplified when responsibility was delegated from Washington to the Denver Land Office to issue patents directly.

Between 1920 and 1960, according to Interior Department figures, patents were granted on some 2,326 oil shale claims, which turned over to private ownership 349,088 acres or nearly 550 square miles of oil shale land. Between 1950 and 1960, patents were issued on nearly 100,000 acres in Colorado alone. Patents were granted on 74,578 acres of Colorado land the claims for which had been canceled for nonperformance of assessment work.* That many thousands of acres of additional oil shale land did not pass from government ownership appears mainly to have been due to the problems speculators and land buyers had in determining the identity of and locating the original owners. A November, 1950, memorandum filed in the Bureau of Land Management Contest evidence from officials involved in Shell Oil's acquisition program states:

Several of the original claimants to parcels in Area "A" [a plot Shell was especially interested in] have died and although considerable time has been spent in trying to trace the claim through a maze of descendants' estates, many of which have not been probated, we were unable in some instances to find anyone with whom to open negotiations. In consequence of these difficulties, we have looked elsewhere for favorable oil shale lands.

It is not precisely clear just when Interior employees began to acquire the first tinges of doubt that there might be something amiss about these routine disposals of public land. The best guess is that everything started one day in 1956 when Scott Pfohl, an attorney in the Regional Solicitor's office in Denver, noticed in examining a group of oil shale claims for which patents had been applied that they had been "quitclaimed" or relinquished to the government by their owner in 1932 because of alleged fraud in their original location. It is the Regional Solicitor's job to analyze such things as a claim's title history before approving the patent application and sending it to the Bureau of Land Management for issuance. Pfohl called Harold D. Roberts, a partner in the Denver law firm of Holme Roberts and Owen, who was representing

---

* After George L. Summers had apparently given up his homestead land, the claims owned by J. D. Freeman, though they had been canceled, were acquired by D. D. Potter and his Federal Shale Oil Co. Patents were granted on the land in 1950.

the applicants, Langdon H. Larwill * and Horace G. Slusser, and said he did not think it would be possible to issue the patent. Roberts replied that another client, an old oil shaler named Rae L. Eaton, the grandson of a former Colorado governor, had been issued a patent on a group of claims several years earlier which were part of the same group and had also been quit-claimed, and that this should be sufficient precedent for a patent on the present claims.

Pfohl had never had too much interest in oil shale, but he decided to check into the patent that had already been granted. He uncovered an interesting chain of events. A man named P. C. Thurmond had staked 17 claims, named "French" Nos. 1–11 and 26–31, on April 5, 1917, and had filed a location in the Garfield County courthouse. But due to a dispute with another prospector, a document had been filed along with the claims indicating that Thurmond had illegally employed "dummy locators," persons whose names were used when the claim was filed so that there could be a full eight-man group for a 160-acre location, but who did not actually have an interest in the claim. Sometimes claimants paid dummy locators a small fee for the use of their names, but often they used people's names without their permission or knowledge. A few years after Thurmond's filing, the French claims were acquired by Rae Eaton, and in 1925 he filed for a patent. A Land Office investigation turned up the fraudulent location circumstances. Patent was refused and the claim was voided. Eaton appealed the decision, but after it was reaffirmed by the Secretary of the Interior, he quit-claimed the land to the government in 1932 so he could receive a refund of his $2.50/acre patenting fee. In July, 1950, Eaton again filed for a patent on the French claims and on August 20, 1951, he received it. Having conveyed partial interest in the claims to members of his family, he then had everyone convey their interests to Eaton Shale Company, which he had created. Standard Oil of California, then engaged in an aggressive land acquisition program, proceeded to acquire Eaton Shale, whose assets included other tracts of shale land beside the French claims, for a reported price of about $4 million. But just from the assignment of the French claims to Eaton Shale, Eaton and his relatives received from Standard of California $957,804.50.

Pfohl turned his findings over to the Justice Departmᵤnt, and on August 5, 1957, just 15 days before the expiration of the six-year statute

* Both Langdon H. Larwill and Harold D. Roberts, whose firm was then called Dines, Dines & Holme, represented claimants at the December, 1926, meeting in Interior Secretary Work's office that was discussed by Ralph Kelley.

of limitations for challenging already issued patents, Justice filed a complaint in U.S. District Court requesting either a cancellation of the patent or a return of the money. The patent had been issued, the complaint stated, through "error, inadvertence, and mistake." In 1960, before the case had gone to trial, Justice accepted Eaton's offer for an out-of-court settlement of $359,176.50, which represents something less than the $600,000 profit it may appear to be since Eaton agreed not to file for a refund of the capital gains tax he had paid. Eaton did, though, retain the patent. The main reason the government settled, apparently, was that before applying for the patent in 1950, Eaton had made a full disclosure of the claims' history to Interior officials and had asked their advice on what he should do. They told him to file his application. "It was a little embarrassing for the government," says A. Edgar Benton of Holme Roberts and Owen, Eaton's attorney. "I think they felt they wouldn't look too well in contending that the application was fraudulently filed. But if the case had come up five years later, I don't think they would have given up so easily."

The Eaton case stirred up a great deal of interest among the other lawyers in the Regional Solicitor's office. The mood had changed from the Byron Mock era and had shifted back to philosophies more in harmony with the days of Ralph Kelley. "A number of us were real novices in the field," says John L. Little, Jr., "and we had no preconceived notions of what the law was and we weren't bothered by tradition. When something like this came along, we hit the books to find out what the law was so that we could present the strongest possible legal case for the government."

There slowly developed a growing skepticism about past policies, and patent applications once processed routinely were painstakingly scrutinized. Accepted legal positions of the past began to fall away, especially the long-revered *Freeman* v. *Summers* decision. For example, in the middle 1950s, John W. Savage, a former Bureau of Mines official and son of oil shaler Harry K. Savage, had applied for patents on the four "Hoffman" claims, and two geological evaluation engineers, Edwin H. Montgomery and Warren Sholes, were dispatched to examine the claims. The claims were atypical in that they consisted of small, roughly 20-acre fragments that had been created by a surveying error. While it is virtually impossible not to find at least a sliver or two of oil shale on a normal 160-acre claim in the Piceance Basin, three of the four Hoffman claims were at such a high elevation that Montgomery and Sholes could not find a trace of oil shale. Montgomery suggested that the claims should be challenged, for even under *Freeman* v. *Summers* they might be invalid. During hearings in 1960, Savage and his attorney, Richard M. Schmidt, Jr., held up a marlstone rock that had been taken from the claim which they admitted contained no oil shale. But they alleged that since the land was part of the Green River Formation, the existence of valuable oil shale could be assumed by geological inference.

This extrapolation of *Freeman* v. *Summers* doctrine caused the Interior lawyers to run to their books, and after some discussion it was decided that in the government's post-hearing brief Interior would challenge the entire validity of *Freeman* v. *Summers*. The trend of subsequent court cases, they felt, had increasingly implied that the mere finding of a few scattered rock samples did not in itself constitute a valid discovery, and recent interpretations of the Prudent Man rule have held that there must be a strong likelihood that the mineral discovered, to be "valuable," must be capable of being marketed commercially at a profit, both at the time of discovery and the time of patent. The fact that an oil shale industry had never come into existence could well be a strong case against present marketability.

Questioned also was the *Virginia-Colorado* decision in 1935. Interior, the lawyers reasoned, may have exceeded its authority in canceling claims for nonperformance of assessment work. It may even have granted patents on the canceled claims. Nevertheless, the cancellations had been made before *Virginia-Colorado* and might be considered to be *res judicata,* literally a thing which has already been settled or decided, and according to legal precedent on the finality of administrative decisions, they should be allowed to stand. The vast majority of the oil shale claimants also had apparently accepted Interior's cancellations. A request for a patent on the claims three decades later might subject the applicants to challenge on the ground of *laches,* or unjustifiable delay in asserting one's rights.

While the validity of *Freeman* v. *Summers* and *Virginia-Colorado* were only a matter of legal interpretation, Rae Eaton's experience with the French claims appeared to have been only one of many instances where patents had been granted on claims which were irrefutably invalid or at least whose validity was extremely dubious. A Grand Valley rancher named Chris C. Dere and his brother Charley, for instance, had filed for 161 claims, covering some 40 square miles of rough terrain, which they alleged to have staked on January 1, 1919, when four or more feet of snow covered the area. At least 14 of these claims covering over 2,200 acres had been patented.

Due to the new outlook, the Regional Solicitor's office by 1959 was disapproving virtually all patent applications. The last oil shale patents, on claims the Solicitor's office had previously approved, were issued on July 6, 1960. In June, 1961, Contest #260, a "test case" employing some of the new legal reasoning, was launched against 35 claims held by Energy Resources Technology Land, Inc., which is controlled by Tell Ertl. In February, 1962, the manager of the Denver Land Office, in the *Union Oil of California et. al.* decision, rejected patents on 257 claims on the grounds that all had been canceled in the early 1930s due to nonperformance of assessment work. Among the patent applicators, besides Union Oil, were John Savage, Tell Ertl, Dr. Charles H. Prien of the Denver Research Institute, and Joseph Juhan.

Assigning the precise responsibility for bringing about this rather unusual shift in bureaucratic procedure is difficult, for the facts of the era have become muddied by controversy. The controversy centers on one of the attorneys in the Regional Solicitor's office named Fred S. March, who later was elevated by some to the status of Ralph Kelley and above as one of the nation's great protectors of the public interest. Indeed just as Kelley had conjured up the specter of Teapot Dome in the late 1920s, March conjured it up in the 1960s. In the end both men, in a series of startlingly parallel incidents, were submerged by the system they tried to combat.

Fred March is a tall, hulking man with large, deep-set eyes and a tight mouth who grew up in a tiny, two-bedroom house in Erie, Colorado, a small town on the flat, poor farmlands in the eastern half of the state. After working his way through the University of Colorado and its law school, he was employed briefly as a land lawyer for an oil company, then joined Interior's Bureau of Reclamation in 1952. Two years later, he was sent to the Regional Solicitor's office as a GS-11 "attorney-advisor" where one of his duties became issuing legal opinions on applications for oil shale patents. March's friends have asserted that he single-handedly turned the government's policy around. According to a letter in April, 1969, by Edwin H. Montgomery, now with the Bureau of Land Management in Washington, nominating March for a Rockefeller Public Service Award, March "did the pioneering analysis which led him to suggest and pursue an urgent and effective attack on existing oil shale land disposal practices. . . . Mr. March had the original idea, made the first proposal, and vigorously pursued . . . major changes in oil shale policy and procedures," such as the challenges of *Freeman* v. *Summers* and *Virginia-Colorado*.

Scott Pfohl died in an airplane accident in 1960 after leaving Interior to work for Sinclair Oil, but the consensus of the other attorneys in the Regional Solicitor's office during this period, such as Robert Mesch, John R. Little, Jr., James Geissinger, and Brian L. Kepford, is that while March played an active role in devising legal strategy, the change was essentially a cooperative effort. All but Geissinger are still working for Interior, however, and Little and Kepford are still in the Regional Solicitor's office. No one denies, though, that March was, as Little says, "a vocal crusader for the office position."

Working late into the night in his tiny, windowless, dimly-lit office, thumbing through yellowed pages of mining law and abstracts of title changes, March became progressively more angered at the way local land speculators had become rich at the public's expense, as he saw it, and how large oil companies were cheaply acquiring title to land containing billions of barrels of oil. To sell land for $2.50 an acre that was worth hundreds of dollars an acre, however technically permissible by the 1872 Mining Law, seemed to him a gross moral wrong. He became convinced the previous granting of patents was not due to legal misinterpretation or lax administration but to a massive conspiracy of rapacious oil companies, unprincipled speculators, and traitorous Interior employees.

His attitudes slowly diverged from those of the other lawyers. He became dissatisfied with the pace of the legal campaign. He wrote long memoranda to his superiors demanding more lawyers, more facilities, and more money to expand the challenges. He was told additional money was not available. He demanded that the government commence legal proceeding against already patented claims before the statute of limitations expired. He was told it was necessary first to lay the legal groundwork on the unpatented claims. When people disagreed with his positions, he suspected that somehow they were part of the conspiracy, and he accused them of selling out, of taking bribes. He wrote long letters to friends in the Justice Department, the FBI, and Congress. When he was told to stop, he threatened to call a news conference to expose the oil shale scandal and others he was convinced Udall was involved in, such as efforts by big oil companies to circumvent acreage limitations in the issuance of oil and gas leases.

He came into hard conflict with B. Palmer King, who took over as

Regional Solicitor in early 1962. King is a thin, somewhat nervous man whose conversation is punctuated with tiny, quick smiles. Though an avowed strong supporter of the new legal effort, he quickly saw March as a distinct threat to his authority. In King's view, a government office can be run efficiently only when complete loyalty, allegiance, and subservience are rendered to the boss. Among his first acts was an attempt to clip the lines of influence March had spread out. March rose up in anger. King tried to calm him and asked him to take a vacation. The situation grew worse as March gathered tightly around him a small group of like-minded supporters, principally Colleen K. Connelly, Thomas M. Stewart, and Albert B. Logan. King told March to improve his "unsatisfactory performance" or face dismissal. One memo sent to March read: "You cause constant turmoil in the office with the other attorneys in that you pronounce strong judgments and make highly critical and insulting statements regarding their work and attitudes"; "you express strong derogatory views with reference to the ability and integrity of certain administrators"; "you infer incompetence or bad faith on the part of everyone who disagrees with your ideas."

Finally, King resolved to get rid of March and his supporters. In 1963, he tried to dispatch Colleen Connelly, an unmarried middle-aged woman, to Albuquerque, but she refused to go, she said, because of her "physical and mental condition." This precipitated long, angry hearings at which March blasted the government's "giveaway" policies on oil shale. In June, 1964, she was fired, and her anxious attempts to win an appeal—she accused Little, King, Kepford, and others of "willfully and maliciously" giving "false and perjured" testimony—were unavailing. King more successfully sent Stewart to the Post Office Department and Logan to the Indian Claims Commission. And he requested permission from Frank J. Barry, the Interior Solicitor in Washington, to fire March for insubordination. Barry, perhaps seeing political liabilities in such a move, instead ordered March to Washington. March's friends assert the intent was to devise a strategy for getting rid of him. Interior officials assert they just wanted to defuse the conflict in Denver and assign March to a new job somewhere else. But as soon as March arrived in April, 1963, he suffered a nervous breakdown and spent several weeks in the hospital.

Frank Barry and Edward Weinberg, his deputy, contacted Michael March, Fred's brother, who works at the Bureau of the Budget, and asked him to come over to discuss his brother's problems. Michael told them that all the pressures and tensions had put a serious strain on

Fred, and he suggested they assign him for a while to a nice quiet post where he could get some rest. Barry and Weinberg said they would do what they could. Not long afterwards, Michael March was ordered overseas for a short trip, but before he left he dropped Weinberg a note marked "PERSONAL" saying that his brother was staying with his, Michael's wife, who was not happy about the arrangement, and he expressed the fear that in his present condition it was possible Fred might do her bodily harm. He asked Weinberg to watch the situation while he was gone.

Weinberg gave the letter to Barry, and the two men, armed with Xeroxed copies, with the "PERSONAL" deleted, started a move to retire March involuntarily for "medical" reasons. When asked about the ethics of using such a letter against its author's brother, Palmer King replies, "You just can't ignore something like that, when somebody makes that kind of statement." After several discussions with March, at which, according to one participant, Barry's display of intense anger rivaled anything ever exhibited by March, Barry subjected March to a series of psychiatric tests. The first doctor, says an Interior official, reported that March was "the worst paranoid I've ever seen." Barry proceeded to press the dismissal through the Civil Service Commission.

A heated struggle followed between the Solicitor's office and an energetic group of March supporters, including a well-known Congressman from an oil-producing state who had gone to college with March. The group spread the word that March was being ousted because of his threats to expose the oil shale giveaways. Once Stewart Udall was summoned over to a meeting at the White House to explain his department's position on oil shale. "That was when I really knew March had gotten through to the Washington level," Udall says. "I decided just to stay out of it and leave it in Barry's hands." Barry, carrying an armful of departmental files, met for an entire morning with David Bell, Director of the Budget, Lee White, special counsel for President Johnson, and others.

The thrust of the lobbying effort mounted by March supporters came to be directed at the Civil Service Commission and its head, John W. Macy, Jr. It was so effective, apparently, that several Civil Service officials became as convinced of the oil shale giveaways as March. A further series of psychiatric examinations of March, who by now had obtained plenty of rest, indicated that he was perfectly normal. Dr. Warren C. Johnson, M.D., wrote the Commission in August, 1964, that, "It is my professional opinion that Mr. March is a well-trained, broadly experienced, conscientious, devoted, professional attorney, who should continue to be given the opportunity to devote his contribution to the government." Interior's retirement request was turned down. "Just when we had it all set," an Interior official complains, "the Civil Service ran out on us." In late 1964, March was transferred to the Portland, Oregon, Regional Solicitor's office, where he was assigned to matters unrelated to oil shale. He soon dispatched an 80-page affidavit to Stewart Udall charging that false testimony had been used against him, but no action was taken on it. "I like dissenters," Udall told me in 1968, "but the problem with March was his fitness to work for the Department, and his emotional problems. He had a great deal of influence on oil shale policy in the Denver office, but then he went off the deep end."

March remained convinced that his enemies at Interior would use the slightest excuse to finish the job of ruining his career, but he nevertheless continued to watch oil shale developments closely. He occasionally sent to Congressmen and other Washington officials long letters and memoranda labeled "ABSOLUTELY CONFIDENTIAL. NOT TO BE QUOTED OR RELEASED TO ANYONE. ANY AND ALL FACTS HEREIN HIGHLIGHTED SHOULD BE CONFIRMED BY YOUR OWN INVESTIGATION." It was his duty, he felt, to warn the nation about what he called "the impending oil shale grab," for he considered his legal victories in Denver to be of only temporary duration, especially since he was no longer on the scene.

Unlike the Ralph Kelley episode, nothing of the switch in philosophy at Interior or of the attempted purge of March appeared in the newspapers—until the appearance of J. R. Freeman, who made the "oil shale giveaway scandal," which he said "dwarfs Teapot Dome by at least a hundred times" and makes it look like a "tea party," his personal crusade. Freeman is a short, lean, wiry man in his middle thirties whose large attentive eyes possess the incandescence of a man who flourishes under the pummels of doubters and nay-sayers. His voice has a kind of dark undertone that can make even a simple exchange of pleasantries seem foreboding. He grew up in DeLeon, Texas, a small town a couple of hundred miles southwest of Fort Worth, and after obtaining a journalism degree from Southern Methodist University in Dallas, he became a reporter for the Dallas *Morning News*. Unable to live on the $76 a week salary, he transferred to the union-represented and higher-salaried composing room where he worked as a printer. His job protected, Freeman, who was a fervid Democrat, began a campaign against the archconservative editorial stance of his employers. When the *News* sup-

ported Nixon over Kennedy in 1960, he collected money from fellow
employees for an ad to demonstrate their dissent with the stand. He
made speeches saying that "the Dallas *Morning News* ought to be called
the Dallas *Morning Assassinator,* assassinator of truth, justice, wisdom.
. . ." But after the assassination of John F. Kennedy in 1963, he became convinced that he "didn't fit," and the following year he, his wife
Elaine, and their two sons, drove north to look for a small newspaper
that might be for sale. He finally came upon the *Farmer & Miner* (circu-
lation 560) located in Frederick, Colorado, a sleepy farm town 30 miles
north of Denver. With $24,000 they had managed to scrape together,
they bought it.

Until that time, the *Farmer & Miner* had not been known for especially aggressive journalism. The former owner had used it to publish
recipes and accounts of visits to relatives in Denver. The Freemans settled down in a trailer adjoining the paper's tiny offices and while Elaine
worked the linotype and handled subscriptions and advertising, "J.R.,"
as he likes to be called, went after crime and corruption. Through several exposés of local shady dealings, he says proudly, "I got rid of the
judge, the city attorney, and a couple of marshals." His blasts at the
miserable condition of Frederick's roads, most of which were unpaved,
produced, with the help of the local congressman, a $173,000 federal
grant. He helped stir up a $3 million suit against the local board of ed-
ucation, culminating in a number of reforms.

An article on Freeman's crusading appeared in *Editor & Publisher*
magazine in early 1965, and it was read with interest by Fred M. Betz,
a Colorado newspaper publisher, and Daniel F. Lynch, a Denver attorney, both of whom were regents of the University of Colorado. They
had been concerned over the university's serious financial problems and
had often thought that if Colorado's oil shale deposits could be developed, they could provide large additions to the state's revenues which
would benefit the university. Freeman seemed just the man to campaign
for such a cause. Betz called Freeman and aroused his interest, and
Freeman began an investigation.

He had published a couple of relatively innocuous editorials on oil
shale when a friend suggested that he contact Colleen K. Connelly.
Freeman located her in Golden and, as he recalls, "I called her from a
drugstore and we talked for five hours, all on one dime." She had saved
a large file of memoranda and other documents from her tribulations at
the Regional Solicitor's office, including a transcript of the 1963 hearings which contained Fred March's detailed testimony on his oil shale

theories. She made all of this material available to Freeman who, on June 9, 1966, published the first of what was to be a 51-part, 200,000-word series on "The Multibillion-Dollar Grab of Oil Shale Lands." There is some confusion about whether Fred March was working with Freeman at this point. Charles Bangert, an investigator with the Senate Antitrust Subcommittee, says that during a visit with Freeman shortly after the series had begun, Freeman acknowledged that March had written the first dozen or so installments. Freeman now denies this, contending that he did not talk with March until sometime after the series had begun. Whatever the case, the bulk of the initial articles were derived from March's investigations and theories.

The thesis, spelled out with fiery vituperation, was simple: All of the pre-1920 oil shale claims were "phony," due to lack of discovery, cancellation for nonperformance of assessment work, fraud in location, and so forth. In granting patents on these claims for $2.50/acre, especially when they could be resold for hundreds of dollars an acre, the Interior Department had perpetrated a gigantic, blatant "giveaway." The only explanation, he concluded, was a conspiracy of Interior officials who were "in cahoots" with "grabber interests," i.e., land speculators and oil companies. He admitted that "overt fraud has not turned up yet dramatically as in the Teapot Dome scandal." Nevertheless, he asked, "Why shouldn't one expect fraud, considering that one 5,000-acre plot of rich oil shale land in Colorado contains more known recoverable oil than the entire state of Texas?" "A full investigation," he asserted, "will reveal that laws have been broken and the public interest has been gouged in an unparalleled fashion."

As had been Ralph Kelley's dilemma, the lack of a specific instance of hanky-panky—even March in all of his digging had not come up with one—was a large obstacle in the way of Freeman's very keen interest in securing national attention for his theories. Another was the fact that the last oil shale patents had been granted in 1960, six years before Freeman began his series. But the longer he continued his research, the more Freeman came to agree with March that the conspiracy still existed and that Interior was "on the verge" of giving away even more shale land. Proof of this, of course, was what had happened to March. Because of March's fear of a renewed attack on his career, Freeman obligingly never mentioned his name in any of his articles, referring only to "horrible instances of attacks and mistreatment of incorruptible public servants." *Ramparts* magazine, the San Francisco muckraking journal, had no such qualms. *Ramparts* became interested in

Freeman's stories in 1966 and sent a team headed by writer Adam Hochschild to review his allegations in detail. Freeman called March in Portland
to ask him to meet with Hochschild. At first March refused, but after an
hour's conversation he told Hochschild that if he flew out to Portland
on a certain date and registered at the Sheraton Motor Lodge he "might
have a visitor." Hochschild flew out, registered, and after sitting in his
room for three days without so much as a phone call, he returned home.

The *Ramparts* story, which was published in May, 1967, nevertheless
emphasized strongly March's experiences as evidence of a continuing,
present-day conspiracy. "If you mention the name of Fred March in
certain government circles in Denver," it said, "people will shake their
heads and mumble something about how 'You can't fight the big
boys.'" A tone of dark sinisterness was pervasive: people "too scared
to talk," mysterious thefts of valuable documents, threats by unknown
men with guns. Colleen Connelly was discussed in detail. *Ramparts* reported that after five of her friends had assembled in her house to listen
to tape recordings she had somehow made of her hearings, "each person
who heard the tapes subsequently received anonymous threatening
phone calls." And that was not all: "Neighbors reported to her that two
men with binoculars were consistently watching her house from a
nearby road. The two men were posing as surveyors. But one day Miss
Connelly stopped and noticed that one of their instruments, a surveyor's
transit theodite, was being held upside down. The next day they were
gone."

To combat the Grabber Conspiracy, Freeman began to deluge representatives of the "grabber interests" with blistering excoriations of their
misconduct and demands for answers to pages of detailed questions on
their giveaway activities. He sent copies of the letters and his *Farmer &
Miner* series to a long mailing list of journalists, congressmen, and
other government officials. Stewart Udall, whom he called "either one
of the most stupid or the most corrupt Secretaries of the Interior in the
history of the nation," was a frequent recipient. "In my opinion," he
wrote in one note, "your dereliction of duty in this matter is tantamount
to fraud." "Perhaps," he suggested in another, "you, as an attorney,
might wish to consider whether you should resign from the law bar if
your trademark is to disregard and violate the laws of the land." When
the Grabber Interests typically failed to respond, he blasted them in the
Farmer & Miner. "It took this writer several months," he said, "to rec-
ognize how deceptive, untrustworthy, and patently dishonest several
high officials of the Interior and Justice Department really are. Several

have lied outright to us on facts which were otherwise commonly known, while others have tried to deny the obvious. These kinds of men, in my opinion, would have fit into Mussolini's and Hitler's operations perfectly, for they would apparently do anything to succeed at the expense of honest people, without concern for ethical conduct, morals, or principles."

Eventually, Freeman became convinced that he had personally come under the surveillance of the Grabber Conspiracy. He told friends his phone had been tapped, that he had received threatening phone calls, that his office had been rifled. One morning in May, 1967, as he was riding near Frederick in his pickup truck, he was shot at from the rear three times. Though hit by glass fragments, he escaped injury. "It was a scare tactic," he told me shortly afterward, pointing out the holes in the back of his truck. "But I think they were willing to run the risk of killing me." An article in the *Farmer & Miner* said that though he had no specific evidence, Freeman "personally considered the oil shale controversy to be tied in because of documents and information in his possession. He explained that of eight people who have obtained or been involved with certain documents, all eight people have either had their lives threatened, or in at least one other case, been shot at." In an editorial Freeman made clear the tactic had been unsuccessful:

The "smell" of death has made us cautious, and absolute fear has made us more vocal to identify key individuals in the executive branch of the federal government who we regard as big oil interest "stooges" on the payroll of American taxpayers. It is with great displeasure that we, as a citizen and newspaper editor, as well as a Democrat, must discharge our duty to the public by identifying none other than the President of the United States as the No. 1 "stooge" for the big oil interests in America.

After the shooting, Freeman started carrying a Smith and Wesson 38 and two large briefcases of his most important confidential files with him at all times. "I sleep with them right beside my bed," he said.

Freeman's articles won him several journalism awards, such as the 1968 Elijah Parish Lovejoy Courage in Journalism Award from Southern Illinois University. In 1968, he was even nominated for the Pulitzer Prize by John Kenneth Galbraith, Senator William Proxmire of Wisconsin, and former Illinois Senator Paul H. Douglas, who has become Freeman's most well-known supporter. During the Senate Interior Committee's oil shale hearings in September, 1967, Douglas introduced Freeman to a group of somewhat startled Senators as a "genuine American hero." The most prestigious national media, however, have not yet taken him seriously, and despite all of the letters and copies of his newspaper series that he has sent around Washington, no federal action has ever been taken on any of his charges.

In early 1968, the financial drain of his oil shale investigations forced him to sell the *Farmer & Miner*. "As the Sun and the *Farmer & Miner* sink slowly in the West," he wrote me, "you may take solace in the fact that this tiny little newspaper, in a place nobody ever heard of, spoke long and loud." Recently he obtained a job as news director for Northeastern Newspapers, Inc., which publishes three small weekly suburban papers in Pennsylvania near Scranton and Philadelphia. He is working on an oil shale book with R. Roger Harkins of Boulder, Colorado, who recently published an exposé of the UFO controversy. The tentative title is *Our Stolen Heritage*.

Fred March, meanwhile, remains embittered and angry, though in 1966, probably as the result of Freeman's articles, he received a promotion and pay raise denied him the ten years he worked in Denver. "My efforts to live by my oath of office and my conscience have been rewarded by horrible punishment," he wrote me not long ago. "I was intimidated and subjected to false charges by my superiors. . . . My personal life was invaded and a potential marriage broken up. . . . Apparently among some Department of the Interior officials exceptional dedication to the public service is equated with mental aberration. . . . What price must a Government employee pay to serve the American people faithfully and well?" "It slowed me down a lot," he said in a later phone conversation. "It's taken a lot out of my will power, and I've lost the tremendous energy that I had. This kind of thing weakens people. It leaves scars." In his nomination of March for the Rockefeller Public Service Award, Edwin H. Montgomery wrote: "To make changes, someone must 'think the impossible thoughts.' And someone must push them. Fred March was such a person."

As engagingly titillating as the idea of a huge giveaway conspiracy may be, it is also, unfortunately, very dubious. In labeling all of the pre-1920 oil shale claims "phony," Freeman includes all of those whose validity is essentially a matter of legal interpretation. He alleges, for example, that to grant patents on claims declared null and void for nonperformance of assessment work is an undeniable giveaway. But he neglects to mention that the 1935 *Virginia-Colorado* decision can be legitimately construed as canceling these declarations and Interior's

Shale Oil Company decision that year, among other things, indicates that was just how Interior officials did interpret it. The same is true with the Department's acceptance of *Freeman* v. *Summers*. One may contend today that it is questionable law but that does not necessarily mean that Interior employees who adhered to its dictates were acting illegally. Freeman has much more of an argument when he discusses cases where patents were granted on claims canceled previously on the grounds of fraud. The Rae Eaton case is convincing evidence that in the issuance of some patents at least, the law was clearly disregarded. It is extremely difficult to believe, nevertheless, that the patenting of oil shale claims until 1960 was the result of a huge conspiracy, necessarily involving several administrations and hundreds of Interior employees and land speculators. Perhaps some of the Grabbers had a close working relationship with a few Interior officials. Some money may even have changed hands, maybe at one of D. D. Potter's parties. But there is no evidence yet to indicate that if such indiscretions occurred they were anything but isolated, individual, local incidents.

A logical general explanation might be, in the words of former Interior Solicitor Edwin Weinberg, "bureaucratic bungling." After 1935, a loose, lenient attitude toward patent applications for oil shale claims became the policy, which was dutifully followed by generations of successive administrators. Many Interior officials undoubtedly held Byron Mock's Western developmental viewpoint, and figured they were aiding the public interest. Putting some of the shale land in private hands might help get it developed, for it was obvious the government was not about to do anything with the land. Why allow a few small legal technicalities to stand in the way of this higher good? And if there was some land speculation going on, it was not the big oil companies who were getting rich, at least not right away, with the shale still lying in the ground and no one around who seemed to know how to develop it. The people making money were the old oil shalers, local boys, who after 20 years of living in near poverty perhaps deserved a few dollars.

(Some of my skepticism about the conspiracy view is the result of my own personal experience. I spent considerable time with Freeman right after the *Ramparts* story was published. I went through his confidential files and documents at length. I even flew out to Portland with Freeman to see March after receiving instructions similar to those received by Adam Hochschild. We were more fortunate, for March soon appeared at our room, and we spent many hours going through the material and files he had brought with him. It was well known among numerous people involved with oil shale, including Interior officials and such Grabber Interests as Tell Ertl and some oil companies, that I was working on an oil shale story for *Life* magazine, of which I was Business Editor. I was certainly in a position to give the oil shale scandal as much national exposure as anyone. Yet in the over three years that I have been researching oil shale, I have never been shot at. None of my files has ever been stolen. To my knowledge, my phone has never been tapped. I have never even received what through even the most extreme leap of the imagination might be termed a threat. As I discuss in the Appendix, there may be some sensitivity about public discussion of oil shale, but so far as I know it has nothing to do with the alleged giveaway conspiracy and it would certainly never manifest itself so blatantly as a threat on anyone's life.)

The theory that Fred March was purged for fighting the "big boys" is also difficult to sustain. Charles Bangert of the Senate Antitrust Subcommittee, after a thorough investigation, told me, "Frankly, I'd love to prove it, but I can't." It is apparently true that some political influence was brought against March. One of the people most upset by March was Tell Ertl who, having just assembled his big blocks of unpatented claims and applied for patents, was caught in the change in patenting policies and was unable to get any of his applications approved. Ertl held repeated conversations with March and felt that March resented his ties to oil companies and his aggressive land acquisitions. "Fred told me several times he was going to keep examining my claims until he found something wrong with them," Ertl says. "I asked him what if he didn't find anything wrong, and he told me he'd just keep examining them." After March had helped file Contest #260 against a group of Ertl's claims, Ertl complained to his friend Congressman Wayne Aspinall, who passed on the complaints to Interior in Washington. "I didn't tell Interior to fire March," Aspinall says. "I just tried to make them get a man who would go along with his superiors." Actually, much greater political pressure was exerted by March's supporters to defeat the involuntary retirement move.

March's impassioned zeal, his impatience to rectify the mammoth atrocities he felt had been committed were what ultimately made his position untenable, particularly since they brought him into stark clashes with Palmer King, the Regional Solicitor, a man whose propensity for enduring dissent by subordinates or assertions of authority that might conceivably come into conflict with his own, is not especially high. King's strong-armed attempts to subdue March could do nothing but make him all the more unyielding and persevering and reinforce his belief that the Grabbers were everywhere. Men like March often thrive in some environments, but not in government bureaucracies. His dismissal was inevitable. The guilt of the Interior Department lies in the rather unethical and underhanded way they went about it.

The most fallacious aspect of Freeman's view of the March episode, however, is the argument that it constitutes persuasive evidence that the Grabbers are still in control and therefore more oil shale land must be "on the verge" of being given away. As the following chapter will describe, it is possible to point to a few incidents during Stewart Udall's term in office which seemed to indicate Grabbers were at work. In fact, however, Udall's protection of the public shale land was as conscientious and scrupulous as any Grabber Fighter could possibly hope. Any Grabber with designs on getting a few chunks of oil shale land must have had to endure eight years of complete frustration. Whether it would have been so if March had not been on the scene is difficult to say. It can with some certainty be said that March's contribution lay in compelling the bureaucracy to dismiss him, for he thus probably created an atmosphere in which the new thrust of Interior's thinking on old oil shale claims was irreversible.

On April 17, 1964, Stewart Udall upheld the 1962 *Union Oil* decision rejecting patents on 257 claims because of their prior cancellation for nonperformance of assessment work.* And on the same day he directed the Bureau of Land Management to conduct a broad legal campaign to challenge the validity of the still outstanding unpatented claims, 407,200 acres in Colorado alone.

The result, filed on September 8, 1964, was what Fowler Hamilton of Cleary, Gottlieb, Steen & Hamilton, attorneys for one of the parties, called "the most extensively and comprehensively presented case in the history of the federal mining laws." It was also by far the most voluminous conglomeration of information about oil shale ever assembled. Hearing transcripts and 1,700 exhibits totaled some 35,000 pages, and the stream of legal briefs that followed the hearings stands over a foot high and weighs 35 pounds.

The case consists of two consolidated contests—Bureau of Land Management Colorado Contest #359 (*U.S.* v. *Frank W. Winegar*) and

---

* Many of these rejections were set aside in 1965 on the grounds that the original claimants or their heirs had not been properly served with notices of Interior's cancellation. But no patents were issued pending possible additional challenges.

Colorado Contest #360 (*U.S.* v. *D. A. Shale, Inc.*)—which challenge the validity of nine oil shale claims covering 1,393 acres for which patent applications had been made. It requests patents be denied and the claims declared null and void. The procedure is semi-judicial. The United States, represented by the Interior Department Solicitor's office, is the contestant, or plaintiff. The claimants are contestees, or defendants. Hearings are held before a BLM Hearing Examiner, who acts as a judge and recommends a decision to the Secretary of the Interior. If the Secretary decides for the contestees, then a patent can be issued. If he decides for the contestants, they may file suit in federal court and, if necessary, appeal all the way to the Supreme Court.

Many of the grounds for challenge apply to the specific claims at issue, which have had an eventful history. Six were located in 1917, and in 1923 they were acquired from an S. D. Crump by a Colorado lawyer and later United States Senator named Karl C. Schuyler. Schuyler was one of the founders of the Midwest Refining Company, and according to the Ralph Kelley articles, two weeks before the expiration of the statute of limitations, Interior Secretary Ray Lyman Wilbur in 1930 refused to approve a challenge against 3,000 acres of shale land for which Schuyler had received a patent despite clear evidence of fraud in location. After spending $100,000 in an unsuccessful attempt to develop a commercial shale oil plant, Schuyler died in 1933, and his oil shale interests, including the claims under contest, passed to his law partner, Eugene D. Millikin. Millikin later married Schuyler's widow and was elected to the United States Senate where he served until 1957, a year before his death. Mrs. Millikin's interests were turned over to D. A. Shale, a family corporation of Schuyler family members, in 1960 in return for stock. The background of the other three claims was much more confusing until they became a goal of Shell Oil's land acquisition program. Shell hired Frank Winegar, a Denver lawyer, to obtain a patent on them, and in 1964 purchased them outright for $30,000. Three additional claims were once part of the contest until Union Oil, who held part interest in them, agreed to quit-claim them to the United States.

As in Contest #260, of which this case is an outgrowth, Interior's chief aim is to win cancellation of the nine claims at issue on grounds broad enough to apply to most if not all of the remaining unpatented claims. The basic argument consists of an all-out assault on the *Freeman* v. *Summers* decision, which the government, in an 100-page appendix, calls "a sad commentary on the operations of the Department of the Interior." The claims did not contain a valid discovery, the government contends, because oil shale was not a "valuable" mineral when the claims were staked and is not a valuable mineral today, for at no time would the hypothetical "Prudent Man" have been "justified in the further expenditure of his labor and means, with a reasonable prospect of success, in developing a valuable mine." (For a patent to be granted, a claim must have been valid when it was located and when patent is applied for.)

The original claimants, says the government, were "individual speculators or promoters who hoped to make a fast dollar through the acquisition and disposal of interests in oil shale claims through the sale of stock in newly formed oil shale corporations." Those men who actually set up producing shale oil retort were "irrational gamblers" who were "mesmerized" by the speculative fever. No Prudent Man would possibly have become involved.

Present-day activity in shale is merely another "speculative bubble," the government claims. A few oil companies are buying up shale claims "not because they have any present value for mining purposes but because they have a speculative or prospective value as a result of competition among oil companies to gain a position in oil shale or by reason of their potential value as a possible source of synthetic liquid fuel at some time in the unknown future." Unassailable proof of this is the complete nonexistence of an operating shale oil industry at any time during the past 50 years as well as the oil industry's present attitude toward oil shale. Nobody is doing anything much, says the government, because everyone knows shale oil production is economically unfeasible:

Estimates of profits (based on assumptions and unknowns) from exploiting Green River shale deposits, and the contested claims in particular, under present facts and conditions, are not sufficient to inspire expenditures of the capital that would be required to commence that exploitation. In fact, they are not even sufficient to warrant the expenditure of time and money in taking the initial step by way of a major research program to obtain the information upon which a rational decision might be made as to whether oil shale is something that might warrant the expenditure of labor and means in actually working the oil shale claims.

The contestees refute this position with lengthy citations of the Interior Department's own statements and activities since 1915. Numerous optimistic forecasts were made of oil shale's great future between 1916 and 1920, they say. The Secretary himself had said in his 1919 Annual Report that: "The country would make no better immediate investment than to give a large appropriation for the development of an economical shale-reducing plant." When applications had been first made for patents, Interior had issued instructions saying that "oil shale has long been recognized as a valuable mineral deposit." Rather than imprudent gamblers, the first group to receive a patent in 1920 included a mayor, a former geologist for the Geological Survey, the governor-elect of Colorado, and the chief chemist for the U.S. Pure Food and Drug Inspection Laboratory in Denver. Interior's encouragement continued through the years, the contestees say, even up to programs announced by Stewart Udall: "The Federal Government has itself publicly and repeatedly stated that the oil shale deposits of the Green River Formation have enormous value and that oil and other products could be produced from the deposits at a profit."

The contestees disagree strongly with the government's interpretation of the present era by citing the ambitious activity of The Oil Shale Corporation, Union Oil of California, Standard of Ohio, plus the frequent optimistic forecasts made by oil men that shale oil was "closely competitive" with crude oil. They quote in detail the oil industry's assertions that it is working hard to come up with an economical shale oil process. They conclude:

It is amid this confident progress toward a commercial industry, with evidence of value more obvious than before, that the Solicitor in 1961 determined to deny the value of the oil shale of the Green River Formation, and this precipitated these contest proceedings. This quixotic insistence that Green River oil shale lacks value at precisely the time when its value has never been greater lends an air of implausibility to the proceedings. . . . (T)he Department of Interior may be cited as a principal evangelist for the view (in every context except this litigation) that oil shale is now valuable and gives promise of becoming still more so.

Much of their supporting argument comes from the fact that Interior consistently adhered to the *Freeman* v. *Summers* decision:

. . . (O)ver more than 40 years . . . hundreds of patents were issued for oil shale claims, each of which necessarily depended upon a finding that oil shale of the Green River Formation was a "valuable mineral" within the meaning of the mining laws. Only since mid-1961 has the Solicitor insisted upon the wholly contrary view. . . . The economic and technological facts to which the Contestant now points are not of course mysteries which have been withheld from those charged with prior administration of the Department of the Interior.

The government does not deny that it has changed its mind. Indeed its case is a strikingly direct repudiation of 40 years of Interior Department practice:

The Contestant . . . admits the standards it is seeking to impose . . . are at variance with the standards that have been applied in the past by some employees of the Department of the Interior insofar as they relate to the adjudication of claims for oil shale lands and the issuance of patents on oil shale claims. It contends, however, that the past actions by some employees of the Department show there was, on their part, a total disregard of even the most basic requirements of the mining laws of the United States.

At other points, the government says these past employees "acted as agents for relinquishing rather than preserving the rights of the United States," that they permitted "an unlawful private appropriation in derogation of the rights of the public." Nevertheless, "the Contestant contends the Department of the Interior need not, indeed must not, perpetuate such errors. . . . (A) history of errors cannot require the United States to part with the title to its lands in violation of the law." The contestees respond that they do not favor perpetuation of errors, but argue merely that 40 years worth of departmental practices supports the contestees' interpretation of the law.

Legal Study of Oil Shale on Public Lands, prepared by the College of Law of the University of Denver under the direction of Professor Gary L. Widman for the Public Land Law Review Commission in April, 1969, and probably the best and most comprehensive analysis ever made, discusses the issue of discovery in detail and especially the recent trend of court thinking on the Prudent Man criterion. It points to a 1968 Supreme Court decision, *United States* v. *Coleman,* involving placer claims for quartzite in the San Bernardino National Forest, which "suggests that the validity of a claim of *any type* will be conditioned on a showing by the claimant that the claimed mineral can be mined, and presently marketed at a profit. It further suggests that this has always been a proper interpretation of the law. If this rule is applied to oil shale claims, it is clear that the claimants face a rigorous evidentiary task."

In response to the "present marketability" trend in recent Court thinking, both the Government and the contestees hassle continually over the economics of shale oil production. The government somewhat grudgingly admits that it may be currently possible to produce shale oil at a profit, but it argues that the profit is not "sufficient" to justify the investment of the Prudent Man; no Prudent Man would put money into shale oil if "the profit is less than can be obtained from comparable risk ventures," and it laboriously compares the oil industry's average rate of return with estimates of the return from a shale oil operation. The contestees contend that producing shale oil is quite profitable, but nevertheless *any* degree of profitability is sufficient to satisfy the law. Moreover, "often large expenditures over a long period of time are necessary after discovery of a valuable mineral deposit to put it on a paying basis."

The trouble with both these arguments, as well as the problem of applying the existing law to them, is that they deal with a situation that does not exist. They assume, as do the mining laws, that if a mineral were valuable enough, the hypothetical Prudent Man would start expending his "labor and means." But in the curious case of oil shale, the more valuable the mineral (to a certain point), the more the truly Prudent Man might be disinclined to become involved, that is if that Prudent Man were the executive of a large oil company. As Chapter Two described in detail, there are numerous reasons why the oil industry is reluctant to expend its labor and means on oil shale that have very little to do with a precise determination of the expected rate of return. Indeed the higher that return, the more of a problem shale oil production could be to its existing crude business. If the oil industry in general begins to expend its labor and means, as Atlantic Richfield appears to be doing now, it will be for many reasons other than a sudden realization that shale oil is "sufficiently" profitable.

This situation leads the two sides in the BLM contests into strange logical traps. The government, at least for the purposes of legal argument, is very happy about the absence of a shale oil industry, but is forced to admit that shale oil production is profitable. It has great difficulty in explaining away such evidence as The Oil Shale Corporation's $47 million worth of activities and high profitability forecasts even from oil executives. The contestees, pleased by the latter, have a hard time explaining the absence of an industry. This absence allows the government to exaggerate its case. It says, arguing that a Prudent Man is justified in developing oil shale is "like arguing that a three-legged horse over the past 50 years has had a reasonable prospect of winning the Kentucky Derby, when in fact no such anomaly has occurred."

Because of the complexity of the issues, which is far greater than can be discussed here, it is difficult to predict how the case will be ultimately resolved. The decision of the hearing examiner, Dent D. Dalby, handed down in April, 1970, follows fairly restricted lines of reasoning. He concluded that despite all the money and effort that has been expended on oil shale, "not one profitable mine has been developed [which] is a compelling reason for concluding that expenditure of money to that end would be imprudent." The only other support of this conclusion in Dalby's relatively modest-sized, 96-page decision was an economic study prepared especially for the case at the request of the Solicitor by the Bureau of Mines. For an $185 million, 60,300-ton plant in 1966, the study predicted a return on equity of 11.3 percent, which Dalby asserted "would be too small to be competitive with petroleum or to attract prudent investment capital." The minimum yield of a typical new investment, he said, would be 14 percent, and for a "nontypical investment such as an oil shale plant, a higher rate of return would be required." *

If this were "a case of first impression," Dalby went on, he would have to find that "oil shale was not a valuable mineral." However, since Interior "has long since decided that oil shale is a valuable mineral" in numerous cases and statements up until the time the present challenges were initiated, he said he must regard those precedents as binding. *Freeman* v. *Summers* was "poorly conceived," he conceded, "but a Hearing Examiner is as much bound by decisions with which he disagrees as by those he endorses." On more technicals grounds— lack of actual physical discovery prior to 1920—Dalby did find that three of the nine claims were null and void.

A decision by the Interior Secretary is not likely until the fall of 1970. Interior lawyers feel confident that Dalby's position on *Freeman* v. *Summers* and other prior Interior rulings will be reversed, since hearings examiners almost always feel more bound by such precedents than Secretaries. If the decision is reversed, the contestees will certainly appeal to the courts. A final ruling by the Supreme Court might not be forthcoming for several years.

There may be a more expeditious resolution to the problem of unpatented claims, however. Three of the parties whose patent applications

---

* Dr. Henry Steele of the University of Houston, in the October 1968 *Natural Resources Journal,* strongly disputed the Bureau of Mines study and contended that if several "cost overestimations" were reduced to more reasonable levels and other adjustments were made, profitability would increase substantially. In general, he concluded, the study was "forensic rather than purely informational."

were denied in the 1964 *Union Oil* decision on the grounds the claims had been previously canceled for nonperformance of assessment work filed suit in District Court seeking to rescind the decision. They were joined by The Oil Shale Corporation, which was not a party to the *Union Oil* decision but has a patent application in for the large block of oil shale claims it acquired or obtained options on from Tell Ertl. Tosco is seeking a declaratory judgment that Interior's old cancellation decisions were invalid.*

The only dispute in *Udall* v. *The Oil Shale Corp. et. al.* then, is whether the old Interior decisions should be allowed to stand. The plaintiffs assert that since *Virginia-Colorado* ruled Interior did not have the right to cancel claims for nonperformance of assessment work, it *never* had that power. In the 1935 *Shale Oil Company* decision, as well as its subsequent patenting practices until 1960, Interior acknowledged that fact. In view of this, the plaintiffs say, nobody should expect them to make an attempt to appeal the old cancellations. The government responds with the *res judicata* and *laches* arguments: even though the cancellation decisions may have been "erroneous," they were within Interior's jurisdiction and they were final. Even if it is determined at a certain time that a decision had been made erroneously, the government asserts, it would be "chaotic" to declare that all other previous decisions "based on that principle, however final, are thereby deprived of effect."

Both the District Court and the U.S. Court of Appeals have ruled against the government. The Appeals Court said it felt "bound by the Supreme Court decision [*Virginia-Colorado*] and by the absence of Congressional action." Since the Secretary of the Interior had acted "without power or authority under the laws" in canceling the old claims, the Court said, his actions were therefore "void and of no legal effect." It added:

however it appears that the problem is a more fundamental one than rule making, and is instead a reluctance by the Secretary to follow the Supreme Court's decision as his predecessors have before, and to seek a new interpretation or to ultimately have the Court modify them.

---

* Due to Tosco's ownership of some 20,202 acres of unpatented claims, its stake in both this case and the BLM Contests is substantial. Though it has no claims formally being challenged at the BLM Contests, it is paying 28 percent of the legal costs. Cleary, Gottlieb, Steen & Hamilton, a prestigious Wall Street law firm for whom Tosco Executive Vice-president Morton M. Winston used to work, is handling the bulk of the legal services in both cases.

The government petitioned for, and was granted in October, 1969, *certiorari* by the Supreme Court. In addition to its other arguments, the government asked the Court to reassess *Virginia-Colorado* and "the implications for the public domain," and whether a claimant should have "a possessory right good against the United States in perpetuity," even though he did nothing to work his claim.

Whether the Supreme Court will decide the case on a strict, narrow interpretation of *Virginia-Colorado* as the Court of Appeals did, or whether it will involve itself in the broader issues as suggested by the government, cannot of course be predicted. Many mining lawyers feel that *Virginia-Colorado* is weak law, a special product of its era, and that the Interior Secretary has more power to administer the public domain than it allowed. The Widman Study for the Public Land Law Review Commission states:

The entire proceeding in *Ickes* [v. *Vriginia-Colorado*] took place during the worst part of the depression. It seems possible that during time of hardship the court might lean toward a decision that would encourage rather than delay the possible development of an industry. . . . At the time of the *Ickes* decision, government intervention into the affairs of business was not looked upon with favor. . . . Today, the role of the Executive as protector of the public interest in the public domain has been modified considerably, and judicial recognition of this expanded role should be anticipated.

Actually, the fact that nothing has been done with the oil shale claims for at least 40 years, for whatever reason, may ultimately be the most damaging legal argument against their continuing validity. The Widman Study comments that the crucial issue is "good faith." "The conferring of rights of claim ownership on a locator," it says, "who complies with the statutory requirements has always been premised on an assumption that the locator was acting in good faith, and wished to develop his claim at the earliest opportunity." In discussing the recent *Coleman* decision, it continues

The present profit requirement was justified by its effect in separating those who entered [the land and staked claims] with the intention to develop mineral deposits from those who entered for other reasons . . . If litigation made it clear that the oil shale locators were subject to a requirement of good faith intent to develop, the fact that the claims have not been developed for some fifty years may create serious obstacles for the claim owners. . . . (F)ailure to develop may not necessarily indicate that the entry was not economically reasonable at the time it was made. But it might be argued that failure to develop, viewed from the perspective of a half century after the original entries were made, does indicate that at some point the locators no longer entertained a good faith intent to develop the claimed land for mineral purposes.

However the *Tosco* case is decided, the government is in the position of standing to win a lot more than it can lose. If it loses, no patents need automatically be issued and the claims will still be subject to the challenges of Contests #359 and #360. "But if we win this one," says Albert Withim, an Interior attorney in Denver who has done much of the legal work on the Interior's challenges, "the whole business of clearing out the old oil shale claims may be substantially over," for almost all of the still outstanding unpatented claims were canceled by those old decisions. The only remaining question would be whether the claimants or their heirs were properly served with notice of these decisions. In a 1965 decision, as was noted, Solicitor Frank Barry overturned many of the denials of patent applications in the 1964 *Union Oil* decision on the grounds that there had not been proper service. Barry's move was opposed by the Regional Solicitor's office who wanted to test the issue in court. But if the government should win the *Tosco* case, it might rescind its strict interpretation of 1965 and deny some applications for patents on claims where service had been questionable. If Interior were upheld, all of the land would again be returned to complete government ownership.

Reading the arguments on both sides of these cases, and talking with the protagonists, one is continually bewildered as to how everybody can be talking about the same body of law, the same rules. One senses, again, the clashing of eras that has been going on since the rise of conservationism under Theodore Roosevelt. The claim owners talk constantly about the 1872 Mining Law, that they hold a valid property right, that just because no one is producing shale oil the government still has no right to step in and take their land away from them. What happened to the old idea about the government encouraging resource development? they ask. *Dis*courage is the word the government now uses, they say. The government is now composed, some of them feel, of socialistic-minded bureaucrats intent upon confiscating private property whenever possible, subduing individual freedom, destroying individual initiative, and expanding the power of the government. Sitting in the living room of an old oil shaler in Rifle, one feels the mood of the frontier, of hardy prospectors out upon the land, wrestling their livelihood from the wilderness.

Back in Denver in the Regional Solicitor's office, and in Washington, Interior employees talk of greedy speculators who are employing high-powered law firms to snatch public land on narrow procedural grounds, land that belongs to *the people,* that great mass which has little defense, save public-interest-minded public servants, against rapacious private minorities who are always pushing for "giveaways." The feelings expressed by these employees have for a long time, particularly since the turn of the century, been held by large segments of the public. It was for this reason, of course, that the Teapot Dome Syndrome first became pervasive. But their widespread existence within the Interior Department itself is a rather more recent occurrence. It is perhaps significant that exposure of both the alleged oil shale scandals originated within the department, not without.

The consequence has been to give the Teapot Dome Syndrome a somewhat altered meaning. Where the motivation once was simple fear of criticism, now many Interior employees in addition sincerely believe it their duty to protect the public resources and prevent giveaways as a matter of deep principle. In a broad sense, this has meant greater attention to the public interest. But in the case of oil shale, it has made it more difficult than ever for Interior to produce a single development program, to do anything but let the shale remain under the ground.

Chapter 6:
Among the cross-pressures
of the Teapot Dome Syndrome,
the oil industry's disinterest and
censure from oil shale's local boosters,
Stewart Udall tries to find a sheltered path.
He succeeds.

"Trying to devise an oil shale development policy has been, I think, one of the most difficult tasks that I have had to undertake in my six years as Secretary," Stewart Udall told a press conference in January, 1967. The difficulty, Udall acknowledges today, was "the tightrope I had to walk" between a multifarious number of individuals and organizations with widely varying ideas on what the government should do about oil shale. Udall, it should be realized, is a very political animal, whose ability to sense the subtle shifts of political pressure is greater than his ability to formulate the most enlightened public policies, whose yearning for political self-preservation is greater than his desire to be immortalized for courageous innovation. (A June, 1969, story in *Fortune* recounted Udall's malleability to the demands of various special interests in his administration of the oil import program and his haphazard granting of "special deals." Fortunately his weakness was of will, not inclination to accept quid pro quos.) But in the case of oil shale, many of the demands on him were mutually exclusive, so that he was bound to be criticized no matter what he did, or even if he did nothing at all. The oil industry was reluctant to cooperate with any federal development program unless it were able to exercise sufficient control over the possibility of future disturbance of its crude business. For the government to grant such terms, however, would inevitably incur the wrath of the Grabber Fighters and probably a lot of other previously uncommitted people who would cry Teapot Dome! and Giveaway!

Practically any kind of development program would also likely provoke the concern and probably the condemnation of the conservation movement, which today possesses political power unequaled since the days of Theodore Roosevelt. Increasingly, the movement's chief interest has become to preserve as much as possible of the nation's few remaining areas of unsullied wilderness, and to maintain as much as possible, the world's "ecology," the delicate balance between living things and their environment which is almost always upset by the pollution and despoilation of an expanding civilization.

Conservationists have had repeated clashes with the Interior Department which though deeply concerned with wilderness preservation has legislative mandates to administer the public lands toward a variety of "multiple uses," such as recreation, water power, flood control, and mineral development. Conservationists have frequently prevented construction of dams and other projects. When an oil well on leased offshore lands near Santa Barbara ruptured in the spring of 1969, blackening beaches and killing wildlife, Interior came under furious attack. When Interior and the Forest Service gave permission to the Walt Disney organization to build a $35 million vacation resort in the Sequoia National Forest 170 miles north of Los Angeles, the Sierra Club, the country's most powerful conservationist organization with 80,000 members, filed suit to stop the proposal. "The only word to describe this hideous project is scandalous," commented *The New York Times,* an avid supporter of conservationist causes. "Has Southern California not been raped, polluted, and desecrated enough already?" A big current battle has been over whether Interior should permit several oil companies to construct a huge pipeline across the frozen Alaskan tundra to bring oil south from the North Slope oil fields. Conservationists maintained the possible benefits to the nation of the new source of oil were minor compared to the great damage that would be caused to the wilderness if the pipeline ruptured, and warned Interior about precipitous action.

Numerous conservationists have expressed their concern over the effects of development of the oil shale. Without adequate "safeguards," Senate committees were told in 1967, development could despoil "one of the truly scenic areas of North America" and "the wintering grounds for the largest migrating deer herd" on the continent. They would certainly become more vocal in their protests if any Secretary of the Interior allowed upon the land huge shovels to strip off surface dirt to expose the shale deposits, towering metal retorts billowing clouds of smoke, ugly heaps of waste slag, miles of pipelines, acres of supporting machinery and employee housing, and all the other conceivable accoutrements of a booming shale oil industry.

It was not possible, however, for Udall simply to do nothing as his predecessors had done. Impatience was growing among the Congressional representatives of the oil shale country, some of whom possess considerable power over the Interior Department. The redoubtable Wayne N. Aspinall heads the House Interior and Insular Affairs Committee through which most Interior legislation must pass. Colorado Senator Gordon Allott, Wyoming Senator Clifford P. Hansen, and Utah Senator Frank E. Moss are all members of the Senate Interior and Insular Affairs Committee. A number of additional Congressmen as well as other public officials have recently acquired the belief that the oil shale land might be the source of incredible federal revenues. The principal missionary of this idea has been former Illinois Senator Paul H. Douglas, who dazzled many eyes over the past few years with his estimate that, as he says in his book *In Our Time,* "Every citizen should know that he now owns from 8,500 to 10,000 barrels of this oil [from shale], worth from 20 to 25 thousand dollars." He actually introduced a bill in 1965 assigning the income to the government from oil shale development to a fund to be used to pay off the national debt. He has waxed eloquent during Senate hearings on how the bonanza can be used in education, housing and a wide expanse of other endeavors for the betterment of the nation. "Riches beyond the dreams of avarice," Douglas remarked to the Senate Antitrust Subcommittee in 1967, after having multiplied $2.77 a barrel times two trillion barrels. (The riches are, of course, somewhat more elusive than Douglas makes out—indeed many of his statements on oil shale show rather pathetically how advancing age (he is 78) has dulled his once sharp and incisive mind. Annual consumption of crude oil is currently about 4½ billion barrels. If *all* of this were supplied by oil from shale, and the profits were so high that the government were able to collect $1 a barrel in royalties, annual federal revenues from oil shale would be less than one-third the present *interest* on the national debt.)

Some of the greatest pressures for shale development came from within Udall's own department, principally from John A. Carver, Jr., Assistant Secretary for Public Lands and later Under Secretary, whose activities and those of a small band of supporters offered J. R. Freeman and the Grabber Fighters periodic bits of evidence that the Giveaway Conspiracy was still flourishing. As Freeman termed it, "The administration which coined the 'Great Society' seems bent on giving away the resources to finance it." Carver, an articulate, persuasive man with the sonorous voice of a preacher, was a member of the Western pro-development school of Interior officials which, though once dominant, had fallen into the minority. He was born in Boise, Idaho, served as attorney for various mining interests in Utah, became administrative assistant for Idaho Senator Frank Church and joined Interior in 1961. One of his first responsibilities was oil shale, and he is candid about where his sympathies lay. "If there is any scandal in oil shale," he told me in 1967, "it is that not one dollar of revenue has come into the U.S. Treasury." Carver resolved to do something about this.

It was thus with some consternation that he learned about the 1962 *Union Oil* decision by the local BLM Land Office denying patents to a large number of claims on the grounds of previous cancellation for nonperformance of assessment work. Carver had already become aware of the activities of Fred March and the other attorneys in the Regional Solicitor's office, and had expressed his displeasure to Frank J. Barry, the Solicitor and Udall's former law partner from Tuscon, Arizona. He told Barry the switch in philosophies in applying the mining laws "questioned the reliability of government decisions. How much can the private citizen depend on what he hears from the government?" All of this litigation did nothing but delay the formulation of a shale development policy. When Congress passed the Mineral Leasing Act, he pointed out, they "expected development." Barry was unmoved and said he fully supported the legal effort.

Carver decided to go outside channels. Within a few days of the *Union Oil* decision, he had an aide draft a proposed bill, a copy of which bears the date February 27, 1962, which would have drastically reduced potential grounds for challenging oil shale claims and would have made the discovery criteria enunciated in *Freeman* v. *Summers* a matter of Congressional law. Carver's aides approached such Senators as Frank E. Moss of Utah and Gale W. McGee of Wyoming to obtain their support. Allies of Fred March within the Interior Department heard about the move, though, and warned associates of the Senators that support of the bill would subject their bosses to the public charge that they were abetting a Teapot-Dome-like giveaway. Eventually Carver found supporters in Colorado Senators Gordon Allott and Peter Dominick. On May 5, 1964, because of "our affirmative duty to create a favorable climate for the early development of a viable oil shale industry," Allott told the Senate, the two Senators, joined by Wallace F. Bennett of Utah, introduced S. 2809 which all but eliminated the power of Interior to nullify the old claims. It said that all owners of oil shale claims could apply for patents and that "neither failure to perform annual assessment work, abandonment, nor the lack of any economically or commercially feasible method of extraction and production of shale oil from oil shale shall constitute a basis for the Secretary to deny issuance of patent. . . ." Applications had to be made before July 1, 1966, though, or the Secretary could move to cancel them on grounds of abandonment.

S. 2809 got nowhere. For one thing, Interior, though refusing to issue an official opinion on the bill, let it be known it was opposed. Consequently, Senate Interior Committee Chairman Henry M. Jackson decided not to support it. Fred March's allies also lobbied against it. Nevertheless, on February 4, 1965, Allott, Dominick and Bennett persisted with S. 1009 which went even further. It expanded the number of grounds Interior could not use against a claim and even allowed anyone whose claim had been canceled or who had been denied a patent to reapply without having the old decision held against them. The deadline was moved up to July 1, 1968. John Carver told the Interior Committee in 1965 that the philosophy of the bill was a "very sound approach." S. 1009 got nowhere, for the same reasons as S. 2809 plus the additional reason of much unfavorable press reaction that Allott and the others were causing a giveaway. J. R. Freeman termed the bill "Allott's Bundles for Billionaires." "Look, deny all the claims, we don't care," says one of Allott aides, "We just wanted to break up the legal logjam and get Interior off dead center. But then, Boom! Everybody says we're in the hip pocket of some oil company." After S. 1009 failed, Allott and the others decided to discontinue the effort.

Carver and his supporters, meanwhile, were employing other tactics. During late 1962 and early 1963, executives of the Shell Oil Company held meetings with John M. Kelly, assistant secretary for mineral resources and former oil man from New Mexico, and other officials. Shell had become interested in *in situ* recovery methods, and had acquired options on some of the long strips of homestead claims along the creek beds in the center of the Piceance Basin. Shell and the Interior people decided that it might be worthwhile for both sides if Shell submitted a request for an oil shale lease on 50,000 acres in the center of the Piceance Basin. "If we could have gotten the land, it would have been a good long-range idea and it would have assured ourselves of a future supply," says W. A. Alexander, retired Shell vice-president for exploration and production in Denver. "They [the Interior officials] encouraged us to take action because they thought it might force some action out of the Department." "Although there is no reason to expect an immediate breakthrough," Alexander wrote Udall on March 19, 1963, requesting the land, "Shell is sufficiently interested to be willing to commit substantial monies for a long-range experimental program providing only that a block of acreage of the type and size justifying such expenditures can be acquired." Shortly thereafter there appeared in Udall's office, carrying additional lease requests, top executives from some of the country's largest oil companies who did not want to miss the opportunity, however remote, to acquire control over the richest portions of the public land. Udall sent the applications to his staff for evaluation and discovered there was enough potential oil in Shell's 50,000-acre request to supply the company, at its present rate of use, for 660 years. Sinclair wanted a 226-year supply. The requests from Union Oil, Humble, Atlantic Richfield, Standard of California, Continental, and Standard of Ohio were somewhat more modest. "Those requests were really scandalous," Udall says today. "Not only would granting them have been fantastically generous, but it would have been risky as all hell."

Udall, who at the time had not thought much about oil shale, ignored the requests, but he decided he had better take some action, or much better, take what might appear to be action. When confronted with a difficult dilemma, the best course, he knew, was to define it, analyze it, study it, and study it some more. In November, 1963, he published in the Federal Register a request for "suggestions from the public at large" on what Interior might do about oil shale. After reading the 200 replies, an Interior official told the 1964 Oil Shale Symposium that: "One fact that stands out is that we must proceed carefully in order to establish the very best basis for the development of our coming oil-shale industry. . . . It was gratifying to see among the responses a recognition of the functional problems that must be solved before development of oil shale public lands can occur."

In June, 1964, Udall appointed an Oil Shale Advisory Board so that "the major public policy questions" on oil shale could be "identified and evaluated." The appointees were a rather diverse lot: Joseph L. Fisher, head of a private study group called Resources for the Future, Inc.; Orlo E. Childs, head of the Colorado School of Mines; John Kenneth Galbraith, the Harvard economist and author; H. Byron Mock, a Salt Lake City Lawyer and former Bureau of Land Management official; Benjamin V. Cohen, a Washington lawyer who served the government under Franklin D. Roosevelt; and Milo Perkins, a Tucson, Arizona, financial consultant who also worked for the New Deal. James M. Gavin, former Army Chief of Staff and head of the Arthur D. Little consulting firm, was also chosen but later dropped out.

The men held four rather heated discussions during which it became obvious they would never reach any type of consensus. The rather paradoxical stance various Interior officials took during the briefings apparently did not assist the process. Byron Mock described it this way in a *Denver Law Journal* article in 1966:

On the one hand we were being told that the resource was of such tremendous value that no one should be allowed to reap the rich harvest of profits from proceeding [to develop the federal shale lands]; while on the other hand we were told that there was no market for the product and that no one could presently or foreseeably treat the oil shale as a valuable mineral deposit for the purpose of discovery under the mining laws.

The February, 1965, "Interim Report" (there never was a final report) contained a 12-page summary of areas of agreement among the board members. Everyone acknowledged the "importance of enlarging knowledge about the resource" and felt that at some point the oil shale should be developed by someone. Then followed 27 pages of separate statements from each of the members. Childs, Mock, and Perkins all seemed to favor commercial leasing of the shale lands, and all were against—in Perkins' case, vehemently, emotionally so—government involvement in research or in "the traditional functions of the private sector of our economy." They generally agreed with Mock who said that "the national interest is best served by immediate commencement of oil shale development." Fisher, as befits his position as chairman, was somewhere in the middle, favoring research and development leases combined with performance requirements to prevent tying up of the land. A specified time thereafter competitive bidding for commercial leases should begin. If private industry response was not satisfactory, he said, the government should conduct its own research program. On the other side were Galbraith and Cohen. Galbraith concluded that the oil industry was "being deterred not by government ownership of other [shale] land, not by fear of what the government may do with these lands, but because of the costs of development and because the further economics of production, as compared with alternative costs of crude oil, are either unclear or unattractive." He went on:

The major oil companies are naturally concerned with protecting their position in the event of the development of an oil shale industry by buying or controlling oil shale acreage. However with one or two exceptions they seem not now inclined to incur substantial development costs to produce shale oil. Certainly for companies with alternative sources of petroleum the economic attraction of oil shale is not high. The incentive to control oil-bearing acreage is thus, for the time being, much greater than the incentive to produce from it.

Galbraith saw no reason for the government to take any more positive view toward oil shale development for there had been "no showing of economic or strategic need for oil from the shale in the present or near future." An immediate leasing program, because of all the unknowns of development and production costs, would be "gravely damaging to the public interest." Galbraith and Cohen felt the government could contract out research, and then when the unknowns were answered, it could make the technology widely available and begin a leasing program. Galbraith was not optimistic that such a course would be followed:

We have been told that the Congress will not authorize appropriations for contracts for the development of oil shale processes. Rather it will succumb to pressure from some oil companies and aspiring lease holders to resist such a course of action. This is an admission that the real interest of those resisting such appropriation is the alienation of the land not the development of the resource. Such obstruction is obviously not a mandate for Executive action that plays into the hands of those who obstruct.

The board's report, in that it tended to obfuscate rather than clarify the issues, was of course very useful to Udall, for the clear implication was that the subject should be given additional study. Additional justification for temporization was being provided by the legal campaign to contest the unpatented oil shale claims. Only about 13 percent of the total shale acreage containing about 5 percent of the oil shale deposit was really at stake, but in his speeches and testimony before Congressional committees, Udall said the legal effort was "a necessary prerequisite to shale development." Unfortunately, this effort would be "time-consuming, vexatious, and difficult." He talked about legal "underbrush" and "thickets" which had to be "cleared out." A very useful by-product of the challenges was the disarming of J. R. Freeman and the Grabber Fighters, who were forced to charge that the challenges were "half-hearted," or that they didn't employ all of the possible legal grounds that had been devised by Fred March. This contention seemed limp, in that it is doubtful, for example, that Ralph Kelley himself could have written the appendix in the government brief on *Freeman* v. *Summers* with more forceful ire.

Freeman and March did have more of a case in criticizing Udall for declining to institute challenges of claims which had already been patented, such as had happened in the Eaton case. The six-year statute of limitations limits the government's ability to "vacate and annul" a patent. However, courts have held that the government may sue to recover the value of a fraudulently issued patent at any time. This possibility was the subject of an angry speech during the 1964 Oil Shale Symposium by Richard M. Schmidt, Jr., a Denver attorney who has represented John Savage and other oil shalers. He warned that if Interior's new legal philosophy were allowed to stand, "all patents heretofore issued . . . could be subject to suit by government to recover the land or the value thereof." Thick political flack would obviously be thrown up if Interior decided to attack all the land under ownership by oil companies, but Interior lawyers have contended that there is sound legal reasoning for not proceeding. "I've been against this until we get some ground rules, some criteria on what a valid claim is, established through departmental channels on the cases already outstanding," says Denver Regional Solicitor Palmer King. "I don't want to go flogging into federal court and have *them* establish the rules and lay down some law that might make it impossible to stop further patenting of the oil shale claims."

John Carver, throughout all of this, was becoming more and more frustrated. "I kept trying to convince Udall of the risks of doing nothing [on oil shale]," he told me, "but he kept hoping the problem would go away." His only accomplishment had been to assist Wayne Aspinall in arranging the reopening of Anvil Points and pushing through the lease agreement, which had been strongly opposed by several Interior officials, especially in the Solicitor's office. Carver and Frank Barry had been continuing their personal struggle over the legal effort. When Senator Allott told Carver at the Senate Interior Committee hearings on oil shale in 1965 that "someone has got to step into the jungle" if all the unresolved legal questions of the cases are going to be resolved, Carver quickly replied, "Yes, sir; and I have been trying to step into it for four years." It was, Carver emphasized repeatedly, "a question requiring Congressional resolution." By 1965, the differences of opinion between Carver and Barry had developed into an intense feud and the two men were literally unable to remain in the same room together. Carver, who had been promoted to Under Secretary directly under Udall, publicly denounced the Solicitor's office in speeches to mining associations and similar groups and he told friends that "the trouble with Interior is that there are too many lawyers."

Udall tolerated the situation, he says, "because I've always felt that the best way to illuminate an issue is by having an adversary situation, with arguments on both sides. In that sense, Carver's developmental point of view was very useful." He understood Barry's viewpoint because "he regarded his role as kind of my watchdog, and he was tougher than hell." Udall's enthusiasm for Carver faded as the altercations intensified and Carver continued developing a friendship with members of Congress, particularly Wayne Aspinall. He also continued his propensity of circumnavigating Interior if he met roadblocks by going through outside channels. "Carver had a lot of friends on the Hill," says a Congressman with a close knowledge of Interior. "He had lots of connections and people regarded him as a real doer, a man who got things done in an environment where actions were usually less than speedy. Some people even saw him as a candidate for Udall's job. Udall talked well, but his critics felt he was not too good on the follow-through."

Fortunately for Udall, the time came for him to make good on a bargain he had made with Wayne Aspinall and Senate Interior Committee Chairman Henry M. Jackson of Washington. When James K. Carr had retired as Under Secretary of the Interior, Aspinall told Udall he wanted Carver moved up to the job. Jackson, however, wanted Charles F. Luce, a Walla Walla, Washington, lawyer and a close friend of Jackson's whom Udall had obligingly appointed in 1961 as administrator of the Bonneville Power Administration. After a brief argument, Jackson agreed to wait if Udall agreed to allow Luce to take the job after Carver had had it for a year or so. Thus, in a three-way round robin during the summer of 1966, Luce became Under Secretary (he later left to become chairman of Consolidated Edison) and Carver was dispatched to the Federal Power Commission to replace David Black who was sent to Bonneville. Carver felt he was being fired. "It was the first time," Carver told me later, "that I ever lost a job."

Following Carver's departure, the demands for action from Colorado, Utah, and Wyoming became louder. Finally on January 27, 1967, seven years after taking office, Udall announced—"with trepidation" he told the Senate Interior Committee—a "tentative" five-step oil shale development program, which consisted of: 1) an attempt to clear title on the

contested land, which was already in progress, and a withdrawal from
metalliferous claiming on oil shale land (the reason for this will be explained in the next chapter); 2) a "blocking up" program whereby owners of scattered tracts of shale land could exchange them for more consolidated tracts of federal land of comparable quality; 3) a system of
"provisional developmental leases" along the lines suggested by Joseph L.
Fisher in the Oil Shale Advisory Board report; 4) a program to en-
list government-industry cooperation to investigate nuclear *in situ* retorting, a move already then in progress by the Project Bronco consortium;
5) a $101 million 10-year research program by the Geological Survey,
the Bureau of Mines, and the Bureau of Land Management which
would entail some unspecified form of "joint" government-private industry activities.

The purposes of the program, as outlined in the departmental press
release, appeared more designed to protect the Interior Secretary than
spur shale development. They were: "encourage competition," "prevent
speculation and windfall profits," "promote mining operation and production practices that are consistent with good conservation management," "encourage fullest use of all known mineral resources," "provide
reasonable revenues to the Federal and State governments."

Interior announced details of the proposed provisional developmental
leases on May 7, 1967, and asked for public comments. (Beyond limiting applicants to one lease and limiting leases to 5,120 acres, the 1920
Mineral Leasing Act gives the Interior Secretary broad discretion in establishing other provisions.) The proposed sequence of events would be:
applicants would request research leases on small sections of a 30,000-
acre track that was being set aside for such purposes—presumably the
sections would be much less than 5,120 acres per applicant—and they
would submit detailed information on their plans for research and eventual commercial production. The Secretary would make his decision on
the applications based on the applicant's "financial and technical capabilities," the "feasibility" of his research plans and so forth. During the
research phase, which could extend over a maximum of ten years, the
lessees would be required to disclose annually all of the technological
data they were gathering, which the Secretary would then "promptly"
release to the public. Rights to all patents and inventions would be
turned over to the government, who would also have total access to all
facilities, records, books, etc. When the lessees completed their research, they would submit a final report containing enough technical
and financial data "to enable any qualified person to carry out the work

performed under the lease," and they would be forced to license any of the other patents they owned if necessary to allow others to carry out the work. At this time, the Secretary would decide whether the shale production process developed by the research was commercially feasible. If he approved, he would grant a commercial lease, though the total acreage might be more or less than the research lease. Meanwhile, the information gathered from all the lessees would allow the government to proceed with a competitive bidding program involving other land. The principal royalties would be closely geared to net income, ranging from 10 to 50 percent depending on how profitable the commercial operations were.

Udall's plan elicited nearly unanimous criticism. J. R. Freeman told the Senate Interior Committee in a long statement that the rules "are designed to continue the transfer of the vast public domain oil shale reserves into the hands of the giant oil companies and other powerful private interests which are and have been robbing the public for many years." They are also designed, he continued, "to foster and to sneak a fraud over on the public. . . ." *The New York Times* in an editorial warned that the government might be pressured thereby into a broad leasing program which might be "a giveaway of unprecedented magnitude." Michigan Senator Philip A. Hart said on the Senate floor that the proposals would "discourage all but the largest integrated oil companies from competition" and thus lead to their "monopolization" of the resource. He warned further that the program required "the most pervasive kind of continuous government regulation with subjective decisions to be made on a regular basis. This could be a stifling barrier to rapid development, and an open invitation to charges of 'giveaway' no matter how proper the decisions." Hart said, though, that he "applauded vigorously" the section about government ownership of patents.

The oil industry did not appear interested, however, in taking advantage of the opportunities these critics said had been presented to it. Udall received comments from 18 private companies, 14 of which were large oil corporations. (Except for Allied Chemical, which was concerned that the regulations might interfere with its leases for trona, a sodium mineral mixed with the Wyoming shale, no large non-oil firm bothered to reply.) All were unhappy about the substantial role of the Interior Secretary through the leasing process. "There is no real assurance," said Phillips Petroleum, "that, after 10 years of effort, the applicant will even receive a shale lease, nor are there any assurances as to the size of the lease which may be awarded." They did not like the acreage limitations which, said Gulf, "has the effect of encouraging a large number of relatively small operations and hence makes economies of scale very difficult."

The most denunciatory language was reserved for comment on the patent and disclosure provisions, which could have a major impact on who the major shale oil producers of the future would be. Udall's proposals had some precedent since early private participants in the atomic energy program had placed patent rights in the public domain and made other technology disclosures. But Amerada Petroleum called Udall's provisions "unrealistic in the extreme." Gulf said they were "particularly onerous." Humble said they would "destroy one of the most powerful competitive forces that could be brought into the oil shale development." Pan-American Petroleum said they were "contrary to the competitive spirit of our form of economy" and that "the temptation would obviously be present for one to wait for somebody else to do the research work." Sun Oil said "this provision completely removes the incentive for carrying out a proprietary research program." Union Oil said that "these provisions are most objectionable . . . There is neither precedent nor moral justification for requiring [a lease applicant] to share its hard-won knowledge with the Department or with others." As a substitution, most oil companies recommended an open competitive bidding program, hopefully without acreage limitations, and with ownership of all technology remaining with the lessees.

Udall defended the patent and disclosure provisions before the Senate Interior Committee in September, 1967, saying they were in accord with the 1963 Government Patent Policy since the leases were in effect a joint government-industry research effort, with the government's donation of shale land being the "equivalent" of "putting [in] actual appropriated money." He added that "the patent policy question bears upon the question of monopoly and antimonopoly" and "any policy whereby the Government made such a large resource available to a few for research purposes without providing that the benefits of the research go to the public would not be a research and development policy, but simply a land disposal policy." In any event, Udall asserted, "some of the spokesmen for some of the companies may say so because they would prefer not to have [the patent provisions], but I do not think it is really going to be a serious impediment." Whether Udall really believed this is not clear. But it would seem inconceivable that the oil industry which already owns substantial shale acreage would agree to spend hundreds of millions of dollars on research on a small publicly owned plot, the results of which would be immediately and freely handed over to potential competitors, including the government, while at the same time being required to face the possibility it might never be given permission to run a commercial shale operation on government land.

In response to the public comments, Udall proceeded to commission another study of oil shale. Originally intended to be of rather modest dimensions, it became a major undertaking by a task force headed by Harry Perry of the Bureau of Mines after the Bureau of the Budget, stirred by talk during the 1967 oil shale hearings in Congress that oil shale might become a huge source of federal revenue, goaded Interior into making an extra effort. *Prospects for Oil Shale Development,* published in May, 1968, was a lengthy analysis of the legal, technological, and economic status of the oil shale problem plus an enumeration of all the multifaceted policy options available to the government. Its principal conclusion was that "the value of the [oil shale] now in place is small and will remain so until new technology has been developed and proved." However, "it is in the Government's interest to encourage industry to make sufficient research expenditures." The report made yet another solicitation for "comments and suggestions."

Udall had hoped that this report might perhaps be his last official act on oil shale. Unfortunately, his Western critics had just about lost all their patience. Governor John A. Love of Colorado and Governor Stanley K. Hathaway of Wyoming—both Republicans—warned Udall that if he did not issue a formal call for leases or take some other genuinely positive step, they would make his lack of action a major local issue in the forthcoming Presidential campaign.

Udall's way out of the dilemma was adroit. On September 10, 1968, he announced details of a formal invitation to bid on "test leases" on three tracts of federal land, a proposal which had been suggested by the May report. The purpose would be "to test the market for oil shale lands to determine whether any company, association, or individual is ready and able to go ahead with development immediately." Leasing was to be by sealed competitive bonus bidding, a procedure commonly used in auctioning offshore lands, and was to be held on December 20, well after election day. According to this method of bidding, royalties and rentals are set beforehand, and bidders offer to pay a one-time bonus above and beyond other payments. The tracts were variously 1,255 acres, 5,120 acres and 5,083 acres containing a variety of geological conditions, such as shale thickness and richness, and presenting "different types of technical challenge." Each, Interior said, contained enough oil to supply a commercial plant long enough for the investment to be amortized. Tract #1, for instance, was estimated to contain 900 million barrels, from which could be produced 125,000 barrels a day for 20 years. Interior officials said they valued the leases at over $80 million.

A number of provisions were more attractive than the 1967 proposals. Royalties were limited to a maximum of 12½ percent, and were not tied to profitability. Lessees would not have to pay anything, including bonuses, to the government for seven years, though the entire bonus would have to be paid even if the lessee for some reason failed to develop the land. Lessees were permitted to retain full right to all inventions and patents, though they would eventually be required to grant licenses to "responsible applicants" at a "reasonable royalty." Numerous uncertainties still existed, however. The terms of the test leases, Interior said, "will not establish precedents for future leasing." The one-lease limitation remained, so that any possessor of a test lease would presumably never have another opportunity to bid for federal shale land.

Reaction from the oil industry was generally almost as hostile to the new proposals as it had been to the old ones. In the survey of major oil companies for this book, not a single one showed even a moderate interest in the test leases. A Phillips Petroleum official said, "The terms were too tough, the financial obligation too great, the risks too high, and the rewards too remote." Most were very critical over the continuing acreage limitations. "The problem was that nobody could ever get another lease," said a Mobil Oil executive. "Interior admitted [in the May, 1968 report] that they had to get that act [the 1920 Mineral Leasing Act] changed, but why should we stick our necks out first? Let them change the Act first. Also, why should I take all the risks being first so all the other companies can take advantage of all my mistakes?" "The terms were ridiculous," said Albert Taylor, a vice-president for Amerada Petroleum. "There was nothing about it that would interest a free-enterprise company." The critics were not pleased either. Wisconsin Senator William Proxmire called the proposal "an oil shale giveaway."

Tosco, Atlantic Richfield, and Shell availed themselves of the opportunity to drill test holes on the tracts, but when the bids were opened on December 20 in the Bureau of Land Management office in Denver, Tract #1 received two bids, one for $249,000 from Tosco and another for $625 from a Fred C. Krafts of Eugene, Oregon. Tract #2 received one bid, $250,000 from Tosco. Tract #3 received no bids. An Interior official termed the bids "absurd" and "ridiculous," and on December 26, all were rejected. "Probably the principal conclusion to be drawn from the lack of bidding interest," a press release quoted Udall as saying, "is that industry apparently doubts that technology has reached the stage where it would be prepared to undertake the large scale investment necessary for commercial development." Despite the failure of the program, the release continued, it had been "a necessary step toward any future program" for it had "cleared away much of the confusion and mystery that have surrounded oil shale for nearly 50 years." Just how the program had managed to accomplish all that wasn't specified. In two letters to Udall, Hein I. Koolsbergen, president of Tosco, sharply criticized Interior's reaction and contended that technology definitely was ready for development. The real problem, he said, was the "uncertainty of the producibility and the lack of knowledge of the subterranean conditions" of the tracts. If it had not been for those conditions, he said, Tosco might have bid $20 million for Tract #2, and would have earned, despite the "unduly high" royalties, a return of 13 percent using the Tosco process. He offered among other things to "materially increase our bid" on Tract #2 if Udall would let Tosco conduct further "geological evaluation." Udall replied that Tosco's proposals were "outside the intended scope" of the test lease program but that they would be "given careful consideration by the incoming administration and the Congress." He thanked Tosco for its interest.

The Grand Junction, Colorado, *Daily Sentinel* commented in an editorial: "Secretary Udall, like the crusading knight of old, is riding away leaving his oil shale bride locked in a chastity belt, a belt constructed in his own bureaucratic blacksmith shop."

"I was delighted to get out of this the way I did," Udall told me recently. "It really silenced the Colorado critics so they could no longer say we were locking the shale up. Of course we weren't surprised there were no real takers. It showed up the people who have been after me for three or four years saying that industry had a keen and lively interest, and that all we had to do was give them a chance. Hickel [Walter J. Hickel, the new Interior Secretary under President Nixon] is going to have the same problem I did. You have to have proposals that are giveaway proof, but on the other hand are attractive enough to be an inducement to private industry. The problem is that industry just is not that interested. If Hickel drafts anything that will attract anyone, a lot of people are going to call it a giveaway. And they may be right."

Interior Secretary Walter J. Hickel, the Alaskan millionaire industrialist and natural resource developer turned model conservationist and public interest protector, has, at least through the middle of 1970, guided his department's policies on oil shale along familiar lines. At his confirmation hearing, Hickel stated that: "I believe we should be prepared to develop oil shale just as soon as it can compete with other sources of liquid fuel in our ecomomy." He took no action, though, until June, 1969, when in response to demands from the oil shale country he and other Interior officials met with Colorado Senator Gordon Allott, Colorado Governor John Love and other local representatives to discuss the situation. Hickel agreed to appoint a task force to make a study.

In early May, 1970, the task force, headed by Reid T. Stone of the Geological Survey, submitted a proposed "prototype" program to Hickel which recommended that 20-year leases be offered on the basis of sealed competitive bonus bids at a fixed royalty rate. Though somewhat similar to Udall's last leasing program, it did have two noteworthy differences. Potential bidders would be permitted to "nominate" to Interior those tracts on which they were interested in bidding. Time periods were stretched out, with a year for core drilling and nomination and an additional year for appraisal of the tracts selected by Interior and of the final lease terms. After two years, Interior would offer six tracts, two from each of the three shale states, at monthly intervals. Since the 5,120-acre limitation on leases was left unchanged, successful bidders would probably be excluded from later sales. Though significant production was not expected for at least ten years, additional tracts might be offered sooner if a reasonable amount of development work had occurred. Except for an escalation of royalties after the fifth year, there appeared to be nothing to prevent the speed of shale development from being totally at the discretion of the lessee.

All indications were that Hickel would approve the proposal. Interior alerted newsmen and sent out preliminary releases. Senators from the oil shale states made announcements hailing the new policy and even ordered that photographs of themselves with Interior officials be prepared to accompany news stories. A subcommittee of the Senate Interior Committee scheduled hearings. But at the last minute Hickel recalled the release and sent a letter to Utah Senator Frank Moss, chairman of the subcommittee, stating that "Our present evaluations of the costs of production from oil shale land lead us to conclude that it would be premature at this time, for economic reasons, to proceed" with the leasing plan. He suggested that the hearings be postponed.

The Western Senators and others rose up in wrath. They demanded to know, among other things, what relevance an a priori judgment of economic feasibility by Interior had for a program involving no expenditures of federal funds and complete assumption of risks by private industry. They wanted to know on what information Interior based this judgment, which seemed at variance with previous Interior judgments. Moss said he would hold hearings anyway and requested that Hickel appear to explain the department's position. Hickel declined to testify at the May 21 hearings, but he sent Under Secretary Fred J. Russell and Assistant Secretary for Mineral Resources Hollis M. Dole. They explained that Hickel, in his letter to Moss, had really been trying to say that the task force had not adequately "delineated" the "economic impact" of protecting the environment, preventing pollution and restoring the land, and had not sufficiently consulted with the states on these questions. Consequently, they went on, private companies could not be expected to produce sufficiently definitive cost data to allow them to decide whether or not to participate. Dole concluded that "it is obligatory at this time that more definitive information be obtained." This, he said, might take from three to six months, but he warned that "my guesses haven't been very good in the past. I've always been six months to a year off."

Since the task force had already considered the environmental issues in some detail, Interior's statement satisfied practically nobody. Moss told me, "I just can't believe it. It's just a cover-up." Speculation on the real reasons for the abrupt turn of events ranged widely, but the most persistent explanation was that, in the words of one high Interior official with a broad acquaintance with departmental shale policy, "Hickel found there was just no pressure from oil companies for moving vigorously on shale. What with all their financial problems on the North Slope and offshore they're just not interested. So he figured: why should we try and go ahead with this thing now? Why not wait awhile? He only cooked up the environmental business when everyone started to scream." Moss said his committee had heard "rumors" that "some of the major oil companies put pressure on the Secretary not to initiate an oil shale development program at this time in view of their great investment in offshore and Alaskan oil." The authoritative *Oil & Gas Journal* lent credence to this view by suggesting that Hickel "was skeptical of industry interest in shale, in view of development opportunities on the North Slope of Alaska."

Some oil executives are said now to feel that the industry overbid on the North Slope leases and substantially underestimated development and production costs.

Earlier in the year, I asked Dole whether it was really possible to devise a shale development program that would both attract private industry and protect the department from political criticism. "Nothing is impossible," he replied. "We intend to find the answers and go forward. We need to investigate all sources of energy as to their possible use." Then he added very quickly, "But there isn't going to be any giveaway, I'll tell you that."

Chapter 7:
A new complication.
It turns out that the oil shale is full of aluminum.
But no sooner is that discovered
than people try to steal it.
Is this yet another Teapot Dome?

Few aspects of the oil shale story are as convoluted and abstruse, or as fascinatingly engaging, as the Great Dawsonite Rush, which involves an alleged snatch of billions or trillions of dollars worth of public resources with so many potential villains that one has the feeling of being caught in the last few minutes of *Witness for the Prosecution*. Lurking beneath the surface, figuratively and literally, is a mineral treasure mixed with the oil shale whose richness, some feel, may dwarf that of the shale itself.

It all begins with an enterprising geological engineer named Irvin Nielsen. Nielsen had worked for Union Oil's oil shale plant during the 1950s, and during November and December of 1963 he was doing some routine core drilling for Cameron Engineers, a Denver consulting firm which had been hired by Marathon Oil to analyze a small, snow-covered plot of land the company held near the center of the Piceance Creek Basin, a portion of the narrow strips of patented homestead claims which follow some of the creek beds in the area. As he brought up the bottom 600 feet of the 2,200-foot core, he was quite surprised to discover, mixed with the lower layer of oil shale, large traces—up to 20 percent of the rock—of nahcolite, a white, crystalline sodium bicarbonate mineral commonly known as baking soda. Nahcolite and other sodium minerals had first been detected in the basin by Tell Ertl in some sampling around the Anvil Points mine. Many geologists had hypothesized that in the prehistoric lakes that had covered the area, an extensive inorganic layer of sodium bicarbonate and sodium chloride had underlain the upper organic layer during the later years of the lakes' existence. Indeed, the Wyoming shale had been found to contain the world's largest deposits of trona, another sodium bicarbonate mineral, which is currently being mined commercially for use in glass manufacturing and other industries. Sodium in the Piceance Basin had always been thought to occur only in small pockets, particularly because the main drilling in the area had been on oil and gas leases. This drilling uses water that dissolves the sodium minerals so they are not noticed. Nielsen's core indicated that the sodium content in the shale might be very high. "I thought maybe we had a real go product," he says, "something that would add to the value of the shale and help its development."

Nielsen immediately told Marathon, Cameron, and several others interested in oil shale, such as officials from The Oil Shale Corporation and John B. Dunn, who works with old oil shaler Joseph Juhan. During January and February of 1964, 17 parties applied to the Denver Bureau of Land Management office for "sodium prospecting permits," which had been authorized by a 1935 modification of President Hoover's 1930 order withdrawing oil shale from further leasing. The permits entitle a party to explore 1,520 acres of public shale land for sodium minerals, and if within two years they are able to make a discovery of a "valuable" sodium mineral and demonstrate that the land it was on was "chiefly valuable therefore" (both terms are extremely vague and subject to even wider ranges of interpretation than "valuable" under the Mining Law) then they will be granted a preference right to a lease to extract the sodium mineral. The background of the modification, however, has been generally interpreted to indicate that a sodium lessee could only mine the sodium if he left the federally owned oil shale undisturbed. In Wyoming, the trona occurs in beds 10 to 14 feet thick that are quite separate from the oil shale.

The 17 permits were issued by BLM effective April 1. The permittees included Marathon, Cameron, Tosco, plus various parties associated with an organization called Wolf Ridge Minerals Corporation, an independent Colorado firm in which a number of local geologists, engineers, and investors held an interest, among them James H. Smith, a former Assistant Secretary of the Navy, Samuel R. Freeman, formerly Assistant Attorney General of Colorado and a partner in the law firm which is counsel to the Colorado Oil and Gas Conservation Commission, John W. Savage, a geological engineer, and Irvin Nielsen. Its president is Arthur Bowes, Jr., a Chicago financier whose family is the largest stockholder in Lily-Tulip Cup Corporation.

As soon as the permits were issued, covering about 20,000 acres right in the center of the basin where Nielsen reasoned the sodium would be found most extensively, Humble Oil lodged a strong legal protest, charging that five of the permits covered land included in its 1963 lease request, which had never been acted upon. Those five permits were quickly canceled, and the BLM office was instructed, additionally, not to issue any more permits to anyone. In an interview with me in 1967, Stewart Udall indicated he had cut off further permit granting because it seemed to him just one more scheme developed by private interests to gain some rights over the shale land. Granting the first permits, he said, "was one of those decisions which had been made without checking with Washington. If it had been called to my attention, none would have been issued." In fact, the permits had been approved by J. Cordell Moore, a John Carver sympathizer who was Assistant Secretary for Mineral Resources. Udall apparently had not been consulted until Humble made its objections.

The remaining 12 permittees were faced with the problem of proving that large amounts of nahcolite constituted a valuable mineral. "Unfortunately, in the beginning we couldn't convince anyone nahcolite was worth a thing in such large quantities," says Nielsen. A portion of the capital being committed to the effort which included drilling several more core holes came from an unusual organization called the JoJo Oil Shale Company, which had been organized in 1958 by John Savage. JoJo is backed by a group of New York investors, many of whom are associated with Loeb Rhoades & Company, a large investment banking house, and who occasionally become involved in such unusual investment opportunities as oil shale. Limited partners of JoJo include Armand G. Erpf, Carl M. Loeb, Jr., and, once again, Huntington Hartford. The JoJo backers supply funds as necessary to John Savage and his wife, Joan, who are general partners.

By September, the permittees still were having only limited success with their analysis of nahcolite, and only a few months remained before the two-year deadline. Then Nielsen received a call from Dr. Charles Milton of the Geological Survey, who had made extensive analyses of the mineralogy of the Green River Formation. Milton had heard about the drilling and wanted to examine some of the cores. Nielsen asked Milton what other potentially useful minerals he had found in the shale during his work, and Milton mentioned one called dawsonite, which he had first detected in 1957. Dawsonite, named for John Williams Dawson, a McGill University professor who discovered it in 1874, is technically sodium aluminum dihydroxy carbonate or $NaAl(OH)_2CO_3$. It had been thought to exist in nature only in small quantities, and is currently manufactured artificially by a patented process to serve as the "active ingredient" in Rolaids, a common antiacid stomach remedy produced by Warner-Lambert. Nielsen asked Milton whether dawsonite might be a source of aluminum. Milton allowed as how it might but, says Nielsen, "he had never given it much thought because he felt it was too hard to extract from the shale." Indeed, dawsonite is tightly intermixed with the shale, probably having been formed from small particles of aluminum-oxide-bearing volcanic ash which at the time of the formation of the Rocky Mountains were deposited into the organic sediment in the prehistoric lakes and acted upon by the saline solution. Though very occasionally tiny, feathery white crystals are visible, the crystals are mostly microscopic and can be detected only by various X-ray diffraction techniques.

Nielsen immediately sent samples of his cores and of cores drilled by Joe Juhan, also a permittee, to the University of Arizona, which sent back a report confirming the presence of large quantities of dawsonite, up to 12 percent of the shale by weight near the bottom, sodium-rich regions of the formation. Later studies were to show that this saline zone ranged over a 40-square mile area in the center of the basin, and that a single square mile might contain, in addition to nearly a billion barrels of shale oil, 126 million tons of soda ash from nahcolite, 100 million tons of sodium chloride and 42 million tons of alumina (Al_2O_3), from which aluminum is easily extracted, which would equal more than one and a half times the United States' entire supply of bauxite, the present source of commercial aluminum. Domestic bauxite reserves are so scanty that the country now brings in some 80 percent of its supply, $200 million-worth annually. Walter E. Heinrichs, Jr., president of Heinrichs Geoexploration Company of Tucson, Arizona, told the 1967 Oil Shale Symposium that "the total potential value of the [Piceance] Basin reserves is approximately tripled by adding the sodium by-products."

Nielsen, Savage, and the others, certain that the key to the undeniable economic feasibility of oil shale exploitation was in hand, began urgent studies on how the combination of sodium minerals could be used commercially. They also started calling large corporations who might be interested. They recruited Kaiser Aluminum & Chemical Corporation, who signed a joint venture agreement, and Advance Ross Corporation, a small TV electronics company which owns 25 percent of Utah Shale Land Corporation, which in turn owns large tracts of shale land in the Uinta Basin in Utah. All of this activity precipitated a great interest locally, and 88 additional sodium prospecting permits covering 200,000 acres were filed with the BLM office, but all were later rejected.

The results of the economic studies indicated that it might be possible to develop all the minerals in the same operation by first retorting the shale at a low temperature, and then leaching out the sodium minerals from the residue. "Of course, there are a lot of unanswered questions which can't be answered without a pilot plant," says Nielsen, "but on paper it looks commercially feasible." Armed with detailed assay reports and core hole analyses, eight of the permittees—including various combinations of Wolf Ridge, Advance Ross, Kaiser Aluminum, John Savage, Irvin Nielsen, and Joseph Juhan—filed requests on April 28, 1966, for leases on 19,109 acres. Though there are some separate, nine-foot-thick beds of nahcolite, everyone knew that it would be impossible to produce the dawsonite commercially without disturbing the oil shale. However, according to an article in 1967 by Jon T. Brown, a former Interior lawyer, in the *University of Colorado Law Review,* neither the mining laws, the Mineral Leasing Act, nor the sodium leasing provisions provides a specific, adequate solution to the Piceance Basin phenomenon of an intimate association of two or more leasing act minerals. In fact, he said, if traditional oil shale leases were issued, where the sodium would be considered just as a part of the shale, "the ultimate resting place of dawsonite and nahcolite might be on an oil shale slag heap." He concluded that "new legislation, specifically designed to deal with this unique situation, may be both desirable and imperative." It was just this sort of solution that the aspirant lessees had in mind.

The scene now shifts to a jovial, leathery-faced prospector from Shawnee, Oklahoma, named Merle I. Zweifel. Zweifel has had a somewhat checkered career—he once spent two years in prison for mail fraud—but he is currently engaged in the legitimate business of staking mining claims. He runs Zweifel International Prospectors, which stakes and files claims for clients, receiving in return grubstake money on the the order of 6¢-10¢ an acre plus a half interest or so in the claims. Typically, he will wait for news of a mineral strike somewhere, and then rush out to stake as many claims as quickly as possible, as close to the area of the strike as possible, in hopes that he might be able to share in the action by selling the claims for a handsome price. According to his promotional literature, "We fly faster, get there sooner, stake faster, than any other firm."

The circumstances of Zweifel's initial involvement with dawsonite are somewhat fuzzy. In *Oil Shale and Offshore,* an occasional newsletter he publishes for clients, he related that he was passing through Rock Springs, Wyoming, in his old pickup truck one day in June of 1965 on his way home from filing some silver claims in Idaho. While stopping for a sandwich, he was approached by a "well-dressed man" and told that there were some minerals contained in the oil shale deposits of the Green River Formation which might be claimed under the Mining Law. He was asked if he were interested. Zweifel said he was, whereupon the man led him into the local BLM office and indicated on some maps just where he should stake.

Zweifel has said that it was at this time that he was told that dawsonite was contained in the oil shale, and that since dawsonite was essentially an *aluminum* mineral, not a sodium mineral, it was subject to location under the Mining Law. This account seems unlikely since the only dawsonite now known to exist in quantity in the Green River Formation is in the Piceance Basin and no one realized the extent of the Piceance dawsonite until Irvin Nielsen's drillings in the fall of 1965. Zweifel now says that while assays from his claims were to show aluminum—traces of aluminum are found throughout the earth's crust, especially in clay—perhaps now that he thinks about it he may have been first interested in staking the shale for selenium, a mineral used in the electronics business. In general, he says, "We were interested in staking the minerals in the shale, whatever they were."

On July 20, Zweifel began what he calls "one of the largest mining claim stake-outs in the United States. We had crews staking right and left." He soon had filed in the Sweetwater County courthouse in Wyoming thousands of claims covering hundreds of thousands of acres of shale land, indicating in the location books that he had found "gold and silver and other valuable minerals." According to Miss Jo Beamer, Zweifel's close associate, "You can find a little gold and silver in almost anything." Zweifel's plan was to file very quietly in Wyoming and Utah before moving onto the more closely-watched Colorado shale lands. Hopefully, by the time anyone else got onto his idea, he would have the entire area plastered.

Unfortunately, the *Oil & Gas Journal* in early 1966 happened to carry a short mention of Zweifel's Wyoming filings, which was read with interest by John B. Dunn, an associate of Joe Juhan, who along with the other sodium permittees was feverishly working on lease applications. "I figured it would be terrible if that guy messed us up and I decided we better do something to protect our permits," says Dunn. He and some of the other permittees decided to go out and file mining claims on the areas covered by their permits and eventual lease requests. Dunn's filings at the Rio Blanco County courthouse in Meeker were noted with interest by Frank G. Cooley, Rio Blanco County Attorney, who went out to file some claims of his own, "It was just a fun sort of thing," Cooley says, "you know like the uranium strike when all the garage mechanics who used to hunt deer on the weekends rushed out and staked claims." This activity precipitated an interview with Keith Miller, a Colorado BLM official, in a local newspaper to the effect that the aluminum content in the dawsonite might well make it locatable instead of leasable, which stirred an even greater crowd of "weekenders," as Cooley calls them, plus a number of owners of unpatented oil shale claims who wanted to protect their land from dawsonite claims.

By the time Zweifel had finished with Wyoming and knocked off a few hundred thousand acres in Utah, he arrived in Colorado to find the entire center of the basin already taken. He did not pause to brood over this ill fortune, however, and went immediately after the remaining shale land. In one of the most aggressive exhibitions of claim filings any Western courthouse has ever seen, Zweifel filed 255 claims on May 2 (beginning with "Aluminum #1"), 485 claims on May 15, 35 claims on May 16, 35 claims on May 18, 42 claims on May 20, and a fantastic 1,279 claims on May 23, for a monthly total of 2,131 claims covering over 250,000 acres or nearly 400 square miles. The total is somewhat imprecise, since in his haste, he appears in a few instances to have filed as many as five claims for a single 160-acre plot of land. Each claim listed eight persons and cost Zweifel about $1.50 to file, since there is a 25¢ fee for each name over two. (It might be worthwhile to remind the reader at this point that the law requires that for a claim to be considered valid, there must have been an actual physical discovery of a valuable mineral on the land and at the very least a posting on the land of a location notice; but until a claim is actually challenged by a subsequent claimant, or the government, the only mandatory act a claimant must perform is to register his claim at the county courthouse.) By early 1967, Zweifel estimates that he had filed some 20,000 claims on close to four million acres of shale land.

Back in Washington, meanwhile, the Interior Department was in something of a turmoil. Stewart Udall had received information about the beginning of the dawsonite rush in early 1966, and many of his assistants recommended that he immediately withdraw the oil shale lands from further claiming under the mining laws. Initial interpretations of the 1920 Mineral Leasing Act had held that land containing a leasable mineral could not be claimed for an accompanying metalliferous mineral not covered by the Leasing Act and still locatable under the mining laws. However the discovery of uranium on public coal lands in the early 1950s had lead to passage of the Multiple Mineral Development Act of 1954 which stated that public lands reserved for leasable minerals could be staked for locatable minerals which occurred separately; the leasing act minerals, however, remained public property and could not be disturbed by the extraction of the locatable mineral. But the problem was determining just how these laws might apply to dawsonite. Was dawsonite a sodium mineral, in which case it could only be leased, meaning all of the dawsonite claims were invalid? Or was it really an aluminum mineral, and therefore locatable? But even if it was locatable, how valid could a claim be for a substance that was so inextricably mixed with a leasable mineral that it obviously could not be extracted without disturbing it?

As if these questions were not complicated enough, there was the further one about the coverage of President Hoover's prior order withdrawing the shale lands. The order had withdrawn the oil shale "and land containing such deposits of oil shale . . . from lease or other disposal." Did this mean that metalliferous claims could not be validly located on oil shale land or didn't it? The 1930 withdrawal was under the authority of the 1910 Pickett Act which specifically excluded metalliferous claiming. As Udall explained to the Senate Interior Committee in 1967: "If we quickly put on another withdrawal order, it would have the appearance we did not believe the 1930 withdrawal order was effective. This is the thing that caused us to pause, because we were not sure that we would be weakening our position rather than strengthening it by putting on another withdrawal." The new withdrawal, further, would have to be non-statutory—though employed occasionally, its legality has been often questioned by legal experts. Udall's legal advisors urged him to withdraw anyway. For one thing, a claimant does not have to specify just what mineral he alleges to have discovered until he is challenged, and once a claim is filed, it can be a tedious job nullifying it no matter how dubious the claim's validity—the problems with the old oil shale claims were proof of this. A further withdrawal might have the practical effect, too, of deterring people from filing additional claims.

Udall, however, did nothing, despite a bombardment of advice throughout the summer and fall of 1966 from numerous Interior officials, including allies of Fred March, and from such private individuals as John Kenneth Galbraith. According to then Deputy Solicitor Edward Weinberg, and several other Interior people, the chief roadblock was John Carver, though Carver now denies it. Weinberg asserts Carver argued loudly against the withdrawals, which he saw as just one more blatant assertion of federal power over the just activities of private industry—it should be understood that at the time it was widely thought large oil companies might be behind the staking—who might be now getting enough land to start developing the oil shale. Weinberg also points out that: "All the Western Senators and Governors look askance at all withdrawals from metalliferous claiming."

Finally on January 27, 1967, after about 97 percent of the Piceance Basin had been covered with new claims, Udall withdrew anyway, as part of his five-point shale development program. From the standpoint of all the new claimants the new withdrawal had the effect of freeing them under the *Virginia-Colorado* decision from having to perform assessment work to protect their claims from subsequent claimants since there could be no challenges from subsequent claims. Assessment work would have cost Zweifel $2 million a year. Reflecting on the incident in 1968, Udall told me that if he could have done it over again, he would have withdrawn earlier. "It would have made us look a little more alert," he said. More recently he admitted to me he felt it was his "chief error" on oil shale.

For J. R. Freeman and the Grabber Fighters, the dawsonite rush and the withdrawal delay were the most scandalous things to happen since Senator Allott's Bundles for Billionaires. As they saw it, Zweifel was only the front man for big oil interests, and dawsonite was, as Freeman termed it, "the modern-day equivalent of the Trojan Horse." He told the Senate Antitrust Subcommittee: "There is a likelihood that the necessary withdrawal order was not issued timely, because those in the Executive Branch favoring disposals of oil shale lands to private interests figured the new mining claims were a possible way with which to let powerful private interests pre-empt the oil shale lands." *Ramparts* said:

"There is a good deal of evidence to suggest that the big rush was largely big companies interested in oil." As evidence of Interior's collusion with Zweifel and the Grabbers, Freeman cited Zweifel's own statement in *Oil Shale and Offshore* that Interior officials had tipped him off on dawsonite and had also made several phone calls from Washington to check on his filing progress. Freeman flew down to Shawnee to question Zweifel on this point, and the transcript of the interview was printed in *Oil Shale and Offshore*. A typical excerpt:

FREEMAN: . . . (L)et me tell you that only two weeks ago I was in the Interior Department offices in Washington, I was just a visitor there. I just mentioned the name Zweifel, and do you know what happened? It was like an electric shock, a bolt of lightning out of the blue sky. Almost literally bells rang, whistles blew, and there was instant bedlam, clerks scurried hither and yon. You ought to try it sometime, you just walk into Interior and announce yourself and see what happens. Just try it sometime. Now will you tell me that you do not influence Interior? Will you tell me that?

ZWEIFEL: Well, according to you newspapermen they probably thought they were going to have a personal interview with the Devil himself. . . .

Zweifel has maintained that he is unassociated with any big oil companies and that he really represents what he called "the Little People of oil shale"—small businessmen and friends, mostly around Oklahoma, who often have grubstaked him, and who "have known and followed Zweifel [in *Oil Shale and Offshore* he always refers to himself in the third person] across North America to the Arctic for many years and this major discovery appears to be their first 'strike' after many failures."

Not all of the people listed in county courthouse records of Zweifel's claims are really "little" but there does not appear to be anyone who is really big. A few oil companies are represented, but the largest seems to be An-Car Oil Company, a $20 million Boston-based oil producer, whose president is John C. Sterge. Along with members of his family, Sterge owns interests in at least 30,000 acres worth of Zweifel's claims on oil shale land. Don H. Peaker and U.S. Beryllium Company, formerly Mile High Oil Company, of which he is head, own at least 130,000 acres. U.S. Beryllium began operations under its new name in 1969 after having been closed down for a few years, and now is engaged in shipping beryllium ore—Zweifel staked some beryllium claims for the firm. It also owns interests in a few Texas oil wells. Warren Shear, a sometime Oklahoma City politician whose wife is a Democratic National Committeewoman, owns some claims but he says that "I've lost all interest in them" because he considers them "worthless."

Most of his grubstakers are people who answered the ads he runs occasionally in *Barron's,* the *Oil & Gas Journal,* and other publications. A recent *Barron's* ad stated:

NEVADA GOLD
Mineral claims staked
for you near new Catlin
gold strike; near new
Cortez gold strike; or
new Howard Hughes' claims,
all for as little as
6¢ an acre.

"It was just a flier with me," says John M. McLendon of Jackson, Mississippi, who owns two radio stations. "I've invested in treasure hunts in Florida, land in the Carribean, always got to have something to satisfy my adventurous spirit, I guess, always like to have something going on. But I don't go buying Cadillacs waiting on the returns." Some of the most eager Zweifel grubstakers are a group of between 20 and 30 people in and around Providence, Rhode Island, who have been following him for years. They are accountants, grocery store owners, farmers, salesmen, and Zweifel occasionally comes up to Providence to tell them about his latest ideas. "What the hell?" says Anthony Nappi, who is in the construction business in Providence. "So I give Merle $150 for 72 claims. Look, I drop $150 in a cocktail lounge and I don't think a thing about it. Sure, it's a shot in the dark. Sure, it's like shootin' crap. So what? Look, I know what people say about Zweifel. Maybe the guy's had a little scrape with the law, and he's done a little time, so he gets a black mark and everybody tries to spread it out. But I'm tellin' you, he's not getting rich. He's not drivin' around in Cadillacs. Sometimes he gets stuck down in Arizona some place, and he's blown a tire, so he wires us and we send him 25 bucks. I mean, he's not getting rich, you know what I mean? Now, sure, O.K., I haven't made a nickel off my claims. But maybe somebody someday's going to come in and knock on my door and say he wants to buy them and I'm going to make something. You know anybody wants to buy them, and you tell them old Nappi, he's ready to sell."

Whether "powerful private interests" are really not carefully hidden amongst these Little People is impossible to know for sure, but it seems extremely implausible that a conspiracy of major oil companies would have selected a man like Merle Zweifel as their agent to gain control of billions of dollars worth of oil shale lands, especially when the validity of the claims Zweifel filed is, as will be seen, rather dubious.

Zweifel, interestingly enough, has also singled out the oil industry as the villain in the Great Dawsonite Rush. In Vol. I, No. 4 (August, 1967) of *Oil Shale and Offshore,* he wrote that:

Zweifel believes that from the very day in July 1965 when he was given the first "tip," that the events were shaped by an international oil conspiracy operating in the legendary Goldfinger style to prevent the United States oil shale deposit from being developed. This conspiracy reportedly has the objective of suppressing the development of United States petroleum reserves so that foreign imported oil will continue to have a ready market in the United States. He believes that the Interior Department has no part in this fantastic conspiracy, but that in some manner, working through minor officials of the Interior Department, the Oil Import Conspiracy has been able to influence or direct the claiming of Zweifel of some 4 million acres of minerals which occur in the oil shale deposits of Colorado, Wyoming, and Utah, and it was their thinking that the Interior Department would challenge the validity of the mining claims which Zweifel had filed, and as a consequence thereof there would be litigation for a generation to come, and of course no oil shale could be produced amid these circumstances.

The Oil Import Conspiracy apparently works in mysterious ways, however. After a flurry of press criticism of his claiming, mostly from newspapers like the *Farmer & Miner* controlled by the Grabber Fighters, Zweifel explained in *Oil Shale and Offshore* that an unidentified "Oil Import Conspiracy lobbyist" had

directed a television and press campaign to influence public opinion against Zweifel. The major theme of the Oil Import Conspiracy lobbyist was that Zweifel was associated with the major oil companies, and that Zweifel did in fact locate his mining claims to deliberately delay the development of the oil shale deposits. This of course Zweifel denies, and he further states that he never at any time has had any dealings with any of these oil companies.

This interpretation may have come to seem a little improbable to Zweifel as concern over his activities grew in Washington. His most vocal critic came to be former Illinois Senator Paul H. Douglas, who devoted most of an entire chapter in his recent book, *In Our Time,* blasting Zweifel and darkly hinting that "there may be big money behind Zweifel." Douglas heatedly questioned Interior Solicitor Frank J. Barry during Senate Interior Committee hearings in September, 1967, on what Interior was doing about Zweifel's claims. The claims were "basically phony and fraudulent," Douglas said, and that his own personal examination of the county courthouse records revealed that Zweifel alleged he had staked many claims during the winter of 1966–67 when many feet of snow lay on the ground. Barry replied that:

We cannot challenge Mr. Zweifel and upset whatever rights he may have simply by an assertion that it was too cold on that day to locate a particular claim. . . . (T)he law permits what Mr. Zweifel is doing. It is not really necessary that a miner or a prospector have a discovery when he locates a claim although the statute says so. . . . Although as against the United States his claim is not valid . . . the courts have held that if an individual locates a mining claim on which he does not have sufficient evidence to qualify for a discovery, he will be protected in his possession against any other person who seeks to displace him. . . . [It will be necessary] to challenge Mr. Zweifel's claims one at a time. . . . We must make a bona fide assertion in a contest within the Department, that there are grounds for justifying its cancellation, and give him notice and an opportunity to be heard so that he will have due process of law.

Later Barry added: "I think I could make this statement without fear of successful contradiction, and that is that Mr. Zweifel, while he may not be a criminal, and I doubt from the evidence that he is, he is a complete nuisance."

Douglas, though, had no time for the subtle nuances of the mining laws. "Must we sit back and allow the public domain, the domain of the people of the United States, to be plundered?" he thundered at Barry. "Just as Zweifel moved on a bold scale to take over the rich treasures of the American people, so I submit we might be bold in their defense!" He went on, "I believe that the Secretary [Stewart Udall] is a splendid public servant * but he will not be Secretary forever. Other and weaker men may follow, as Albert Fall followed Franklin D. Lane. . . ." Other witnesses at the hearings recommended a "legislative taking" of Zweifel's claims, a swift, surgical stroke which would chop off the Big Money right at the roots. After the hearings, Douglas announced formation of the Public Resources Association, with John

---

* Douglas, while admiring everything else about J. R. Freeman's Giveaway thesis has been very critical of Freeman's views on Udall, who is a good friend of Douglas.

Kenneth Galbraith as his honorary cochairman, to serve as an "Oil Shale Watchdog Committee." Colorado Senator Peter H. Dominick commented that, "we need this organization like we need a hole in the head." (Having failed to raise $115,000 for the "first year's research," the PRA is now more or less defunct.)

In the face of all this hostility, Zweifel in Vol. 2, No. 2 of *The International Prospector* (a temporary replacement for *Oil Shale and Off-Shore*) revised the assignment of villains:

The major oil companies holding old and defunct titles to the oil shale placer claims were furious at Zweifel and his grubstakes for filing a challenge to their claims. Their first action was to employ an astute and clever lobbyist to influence public opinion against Zweifel; someone who had an honest record in Congress and would not be suspect of scheming and conniving against little people everywhere. The ex-Senator from Illinois seemed to fill this requirement perfectly. . . . Senators Proxmire and Douglas (Democrats of course) hypocritically speak of organizing the oil shale "conferences" [through the Public Resources Association] to protect the public from a "Teapot Dome Giveaway," and loudly proclaim that their motives are unselfish and in the public interest. But in their organized efforts to cancel out the mineral claims of the little people they are discreetly cloaking the birth of a hundred Teapot Dome scandals.

That Paul Douglas, who spent years combating efforts of large oil companies and state governments to secure rights to offshore oil lands was now really a conniving employee of the oil industry may have proved a little difficult for some of Zweifel's grubstakers, who include many Democrats, to assimilate. By September, 1968, Vol. 2, No. 9, the evil force trying to crush the Little People had assumed another guise: "Left wing and socialist groups of the U.S. Congress." Commenting on a trip to Shawnee by some Interiors lawyers to secure information on his claims Zweifel wrote:

We know that this extended foray into Rebel Country by Interior Department Attorneys is a pre-election political move by the administration to placate liberal elements in the U.S. Congress that have long advocated nationalization of the oil shale deposits. It will indeed satisfy the liberal elements of the Senator Douglas group that Zweifel and his cohorts be drawn and quartered for daring to advocate private enterprise development of the oil shale deposits.

In a recent interview, I asked Zweifel to explain to me just who he felt was out to get him and the Little People. He said that he really wasn't sure, and he admitted openly that his views had changed. "To tell you the truth," he said, "I've thought a lot of things. This thing sure is complex and confusing, and sometimes you just can't be sure *what's* going on."

A good deal less murky is what Zweifel's ultimate intentions are. The government may still own the oil shale, but Zweifel figures that he owns the dawsonite and other minerals which are intermixed with the shale. "If the government won't give us permission to mine oil shale along with the minerals we staked," Zweifel said in 1968, "then we won't give them permission to touch our minerals. Our right to our minerals is statutory, and if they disturb our minerals, we'll get a court injunction to stop them." According to Charles Nesbitt, who while attorney general for Oklahoma was approached by Warren Shear about joining the grubstakers for Zweifel's venture and who met with Zweifel, "Zweifel's out to shake somebody down, anyway he can. He's interested in the claims for their nuisance value. He figures, for instance, it would be to the advantage of any oil company leasing the shale land to buy him out and avoid a long and messy court case," Zweifel is willing to settle with the government, he says. In a letter to Stewart Udall in June, 1968, he offered to relinquish his rights for ⅔ of the gross federal income from royalties and the sale of oil shale leases. In October, he offered to sell his claims for a mere $200,000/acre, which, assuming apparently that extraction of the minerals will be 100 percent profit, he points out is "roughly one-eighth of the actual value of the mineral claims." If this price prevailed for all of Zweifel's claims, the Little People would gross some $800 million, which is enough to make a lot of little people very big. The only nibble so far, Zweifel says, was an offer by Cleveland Cliffs, partner in the Colony venture with Tosco, to buy some claims for $150 an acre if legislation were passed allowing mining of the dawsonite along with the shale.

In August, 1969, the government and Zweifel began what Zweifel calls "a legal battle for control of the world's richest mineral resource." Contest #441 by the Bureau of Land Management challenges the 2,911 claims Zweifel alleges he located in Colorado on the grounds that they were "not located in accordance with the mining laws" and that "no discovery of valuable, locatable mineral deposit within the meaning of the mining laws has been made within the limits of the claims, or any of the claims." Zweifel will be required to show that on every single one of his claims he made a specific discovery of a valuable mineral at a specific location. Hearings were scheduled to begin in June, 1970.

If Zweifel asserts that the valuable mineral he discovered was dawsonite, then a chief legal issue may be whether dawsonite is locatable or leasable. In an apparent attempt to quell criticism that it was not doing anything about Zweifel, Interior announced in May, 1968, that it had determined dawsonite was leasable and therefore Zweifel's claims were "wiped out," which drew immediate comment from *The New York Times* that the Department "deserves praise for this defense of the public interest." Of course as Frank Barry's testimony before the Senate Interior Committee indicates, a mining claim cannot be "wiped out" by such a determination. Still, most mining law experts feel that if the leasable *v.* locatable issue comes up in proper contest proceedings, the government's view will probably be upheld. Previous decisions on the status of so-called "double salts" such as potassium aluminum sulfide have held them to be leasable. Moreover the courts tend to abide by Interior's judgments in such matters because of its special competence.*

Much of the government's case against Zweifel, though, is likely to focus on the techniques he used to stake his claims. Zweifel's statements in a September, 1968, deposition are not likely to inspire confidence among his grubstakers. He testified that without surveying instruments of any kind or any guide other than a compass and a gas station road map, he staked his claims by driving his pickup truck back and forth across the Piceance Basin—which is filled with numerous canyons and which contains only a few rough, winding dirt roads—and, judging distances by his odometer he periodically hopped out of his truck, pounded in a location notice, picked up a few rock chunks from the surface of the ground as a "discovery" sample, and "threw [them] in the truck." He freely admitted that he did not usually make an immediate attempt to tie a particular rock sample to a particular claim. Only later in his hotel, or campsite, or "when I felt like it, if I got tired" did he mark the rocks with a yellow pencil. This would imply a rather difficult sorting problem since "I usually picked up maybe a hundred or two hundred samples" in a day. Zweifel took the rocks back to Shawnee and actually sold most of them—his backyard is full of rock samples which he advertises in brochures he sends out through the mail—as "samples from the world's greatest potential energy source." Zweifel presented

---

* In February, 1970, Interior announced it had convinced holders of 198 claims, including Atlantic Richfield, Marathon Oil, Tosco, and Cameron Engineers to relinquish them to the government. Frank Cooley and some of the other "weekenders" asserted they intend to hold on to their claims just in case they should turn out to be valid.

several assay reports showing various degrees of mineralization * but he later admitted that in all but two or three cases the samples for which the assays were made had been obtained from the area of his claims after the January, 1967, withdrawal. Zweifel excuses this somewhat haphazard procedure with the contention that since the land was withdrawn he did not have to worry too much about some of the technicalities because he won't have to defend his claim against subsequent claimants. He said he couldn't do much more than pick up loose rocks because he might have disturbed the government's oil shale and "they might have been tickled to death to catch me with a chunk of that oil shale and said 'Boy, we have got you now.' . . . They might have throwed me in jail over it." Despite all of this, unless he can show he had made a valid discovery *before* the withdrawal, his claims will be void.

There is a fair amount of evidence to indicate that Zweifel really did nothing in most cases but file "paper locations" at the county courthouses. Charles Nesbitt told me, "Hell, Zweifel said he'd actually never been out on the land, that you really don't have to." John Franks, head of the Rock Springs, Wyoming, BLM office, says an investigation of a million or so acres of land a month after Zweifel alleged he had staked claims on them showed no evidence whatsoever of location notices or any other sign he had actually physically staked mining claims. Several of the men who filed dawsonite claims in the Piceance Basin say they never saw either Zweifel or, except the filings in the county courthouse, any sign he had been there. Zweifel says if his location notices are gone, "It must be because deerhunters used them for firewood."

The burden of proving Zweifel's claims are valid is in the hands of Clement T. Cooper, a Washington, D.C., lawyer with his own practice who met Zweifel after answering one of his ads in 1960. Along with his wife and other relatives, Cooper holds an interest in 50,000 acres worth of Zweifel's claims on the oil shale lands. He admits that all the evidence indicating Zweifel filed only paper locations "starts you thinking." However, he says, "Merle is a likable, sincere guy, real gentle." Regardless of what Zweifel really did on the land out in the Piceance Basin and elsewhere, Cooper still feels he will do all right in court.

* Zweifel alleges that his assays show aluminum in the Colorado claims. But since the dawsonite lies almost entirely in the deeply buried sections of the Piceance Basin and since his claims are mostly around the rims, it is likely any aluminum traces are the same small traces that can be found in almost any section of the earth's crust.

"The government is going to have a rough time with me," he says. "I'm going to make things so confusing and so messy, that either I am going to win or nobody is going to win. I'm strong on this. I feel strong on this. See those books over there? Well, every night I'm studying until one or two in the morning, and I'm getting stronger because you got to be strong to fight the government. Millions of dollars are involved, you see, and that is what motivates me. The way I figure it, in America, you get a shot at a thing, and when that thing comes along, you got to take it." He pauses and smiles. "But then of course somebody . . . might . . . come . . . along and relieve me of the responsibility, with a little quid pro quo, a little cash, you know what I mean? Like if one of those oil companies wants to buy me out, well, I might just go along, you know?"

Barring such a happy circumstance, Merle Zweifel appears no less determined. The Little People, he contends, will triumph. "We'll fight 'em for 25 years if we have to," he says. "We'll fight 'em down to the very last acre."

While Merle Zweifel has run into a lot of opposition in his attempt to carry off a four-million-acre land grab, the requesters of the sodium preference leases have had no easier a time trying to obtain leases on a mere 20,000 acres. Throughout 1966, 1967, and 1968, representatives of Wolf Ridge Minerals Corporation, Kaiser Aluminum, Advance Ross, and various affiliated interests met numerous times with Interior Department officials without avail. Considerable legal effort was mustered by Advance Ross who, in a display of bipartisanship, retained the New York law firm of Nixon Mudge Ross Guthrie Alexander & Mitchell (Nixon and Mitchell were later lost to the new administration) and the Washington firm of Chapman, DiSalle, and Friedman (Oscar L. Chapman * was Interior Secretary under President Truman, Michael DiSalle was Democratic governor of Ohio, Martin L. Friedman is general counsel for the Democratic National Committee).

In early 1968, Edward Weinberg, who took over as Solicitor from Frank Barry, prepared an opinion rejecting four of the lease applications. "To permit the extraction of minerals which are in themselves a constituent of the oil shale rock or which requires the removal of signif-

---

* It was Oscar Chapman who, as Assistant Interior Secretary, in 1934 rejected a request for a sodium prospecting permit, an appeal of which precipitated the 1935 modification of the Hoover withdrawal order authorizing the permits and the leases.

icant quantities of the surrounding rock, itself oil shale," it said, "would obviously affect the purpose of the oil shale withdrawal order, which is to reserve the oil shale from mining and disposition." It did acknowledge that "there is some question" whether a bed of white nahcolite revealed in the permit applications might be valuable, and the applicants were allowed to submit further evidence on this point. The opinion added in a footnote that the "interlocking ownership" of the various lease applicants should be investigated to see if there had been a "violation of the sodium acreage limitation."

As soon as the applicants heard about this, Martin Friedman stormed over to the Solicitor's office, "bitching like hell," according to one Interior lawyer. After some discussion, Weinberg agreed to modify his decision. In May, 1968, he ruled that the applications be sent back to the Bureau of Land Management for hearings before a Hearing Examiner at which the government and the applicants could present their cases. In October, 1969, Senator Philip A. Hart announced that he feared a "giveaway" because he had heard that Interior was about to issue the leases without holding hearings which, he wrote Interior Secretary Walter Hickel, "would deny the government the opportunity to present evidence which the Department previously thought was necessary for the proper disposition of the matter." He added he was worried that if the leases were issued "the companies could acquire the valuable oil shale deposits by paying only the minimum royalties and without having to offer competitive bids." Hickel swiftly issued a statement saying that "no decision had been made" and that "Nobody is going to give away oil shale or any other valuable resources in our custody as long as I am Secretary of the Interior."

"Everyone's been talking about political pressure in this thing," says Karl Ranous of Van Cise, Freeman [Samuel R. Freeman of Wolf Ridge], Tooley & Eason, a Denver law firm representing Wolf Ridge. "But I do know that it's been four years and we haven't gotten our leases yet."

Kaiser Aluminum, meanwhile, eventually decided to give up its lease requests. "It just became a political football," a Kaiser official says. Kaiser studied the problems of getting aluminum out of the shale oil, but discovered that the alumina content of shale is only 4 percent against 50 percent in bauxite ore. "It's really very intriguing, a great conversation piece," said the official, "but the economics just didn't look very good," though he feels that eventually some form of *in situ* retorting coupled with a leaching process to extract the alumina might prove feasible.* He pointed out that one factor in Kaiser's decision is its 40-year supply of foreign bauxite reserves. A March, 1967, *Forbes* article quoted Leo M. Harvey, chairman of Harvey Aluminum, as saying that a barrier to large-scale aluminum industry development of the Piceance Basin was its $1 billion worldwide investment in traditional aluminum extraction from bauxite by the Bayer process. "You just can't drop all that and go somewhere else," Harvey said.

Seemingly unconcerned about all their difficulties, the remaining lease applicants have been proselytizing professional engineering societies about what John Savage calls "the world's greatest treasure chest of chemicals," which he frequently points out never would have been revealed if the prospecting permits had not been issued. Irvin Nielsen told the American Institute of Mining and Metallurgical and Petroleum Engineers in May, 1968, that billions of tons of sodium plus other minerals extant in the area can serve as "raw materials for an incredible inorganic chemical, glass, aluminum, building material, and cement industry. . . . The chemical and other products that can be made with these resources are simply too numerous to comprehend. However, here is a chance for inventive chemical engineers to have a field day." For instance, nahcolite (the Piceance Creek Basin contains the world's only known major deposit) could be a valuable air pollution control agent, for it easily reacts with polluting sulphur compounds emitted in stack gas; Wolf Ridge is working actively with a Department of Health, Education, and Welfare air pollution research group. If a mixture of oil shale and nahcolite were mined and produced, he predicted, the profit before taxes, royalties, and interest would be 91¢ per pound of mined rock against only 13¢ for oil shale production alone.

Somewhat ironically, the role of oil shale in this glorious vision of the future is quite minor. (One reason may be the applicants' efforts to prove that the shale lands are "chiefly valuable" for sodium.) In one of the proposed multi-level "chemical-metallurgical complexes" which will produce glass, fertilizers, aluminum, and all manner of other products, the shale oil that is produced is almost completely used up right on the plant site to fuel the complex, especially to generate the large quantities of electrical energy needed to obtain aluminum. Wolf Ridge has told Interior it would even be willing to give all the shale to the government

---

* Tosco announced it had developed an alumina recovery process in cooperation with American Metal Climax, but that it was commercially feasible only in conjunction with shale oil production.

if the government would allow them to dig up the shale to get at the nahcolite and dawsonite.

Irvin Nielsen, in fact, vehemently attacks the very term, "oil shale," which he calls an "atrocious misnomer." The mineral, he asserts, should be given "a unique name." In one speech he said:

The name "panaceaite" was once proposed as an allusion to the many problems that could be partially solved by a multi-purpose industry which would exploit its values. When you think of the national problems that could be aided by this new industry such as air pollution, balance of payments, strategic minerals, unemployment, etc., the name "panaceaite" no longer seems ridiculous.

PART FOUR

"Economic history teaches that we must take the long-range view of the industrial evolutionary process. The historic antitrust case of 1911 resulting in the dissolution of the Standard Oil trust played a central role in promoting competitive behavior in the succeeding half century. Similarly, the manner of disposition of our vast oil shale resources almost certainly will have a profound impact on the industrial organization of the petroleum and broader energy industry over the next half century."

—Willard F. Mueller,
Director, Bureau of Economics,
Federal Trade Commission,
May 5, 1967

Chapter 8:
Clearly, it all adds up
to a massive impasse.
Is the bonanza destined
always to remain elusive?

The oil industry seems happy enough without it. Government officials are scared to touch it. The general public has never heard about it. Except for a couple of members with vested interests, Congress couldn't care less about it. The obvious question at this point, after seven chapters worth of political, economic, legal, technological, and psychological complexities whose resolution would challenge even the most venturesome and resourceful administrator: Do we really need oil shale? Why not just let it sit in the ground?

One of the many unfortunate results of the widespread assumption that the oil industry is the most logical developer of the oil shale has been the equally widespread assumption that the oil industry's very special, private view of oil shale is synonymous with the public interest. In particular, the chief reason why oil companies have been accumulating oil shale land is to provide themselves with shale reserves if the nation should run short of crude reserves. Practically the only companies to show appreciable interest in shale development have been crude poor and fearful, should such shortages arise, of being caught in a profit squeeze or, worse, without a supply of crude for their refineries. The North Slope discoveries, oil men say, push further into the future the time when shale oil will be needed. The government, accepting completely this line of reasoning, has generally taken positive action toward oil shale development, such as the construction of Anvil Points, only when serious crude deficiencies for the country were forecast.

It is possible to make a fairly convincing case that eventually the

United States will have to supplement its domestic crude production with increasing amounts of oil from some outside source. The argument runs thus:

1) Current demand for crude oil is increasing more than 3 percent a year and shows few immediate signs of letting up.

2) The pace of crude oil discoveries has been irregularly declining since the 1930s. Exploration and drilling activity has been also declining at an increasing rate. Thus while domestic crude production in 1969 totaled 3.2 billion barrels, additions to reserves totaled only 2.1 billion barrels. Proved reserves at the end of 1969 were 29.6 billion barrels, down 3.5 percent from 1968, representing a 9-year supply against a 13-year supply in 1960.

3) If these trends continue, reserves will eventually be depleted and where once supply exceeded demand, demand will exceed supply by ever increasing amounts. The *Oil & Gas Journal* has predicted that excess capacity in Texas and Oklahoma will vanish sometime in the mid-1970s. A Humble Oil official has said that by 1985, 85 percent of the country's requirements will have to come from reserves not yet discovered, and that even if all domestic oil fields are producing at full capacity, production will equal only 82 percent of demand.

Similar arguments have been advanced for the past century, and it has always seemed that just when the state of the crude supply looks bleakest, a few more huge oil fields are uncovered. This is why present assertions of impending shortages are the subject of intense debate. Though very unlikely, the North Slope just might contain 50 billion barrels, which if production costs are not prohibitive, could free the country from worry for many years. Some people believe that new methods of secondary recovery of existing oil fields will be developed, for at present often 70 percent of a reservoir's oil is unrecoverable under present technology. And if the country were willing to accept the possible political risks and the unfavorable balance of payments, we could rely on cheap oil in proven Middle Eastern fields to supply the bulk of United States needs for decades.

The view is also widely held that the internal combustion engine, gasoline for which accounts for half the nation's refinery output, will soon become obsolete, probably because of public concern over air pollution. It is true that the automobile now causes between 40 and 60 percent of the country's air pollution, but one should not underestimate the efforts the automotive and petroleum industries are capable of mounting to protect their huge capital investment in established ways of doing business by "cleaning up" the IC engine rather than being forced to switch to a new power source such as steam or electricity. For a while the two industries engaged in what one writer called an "Alphonse-and-Gaston routine," wherein the car makers said they would produce cars that could burn non-polluting gasoline when the oil men came up with the gasoline and the oil men said they would make the gasoline as soon as Detroit started producing engines to burn it. This posing was dropped when it was realized just how serious and widespread environmental concern had become, and both sides started taking positive action, such as removal from gasoline of tetraethyl lead, an octane-enhancing additive which clogs anti-pollutant devices. Henry Ford II, chairman of the Ford Motor Company, said that environmental pollution was "by far the most important problem" facing his industry in this decade. It seems certain that within 3 or 4 years, automobiles will emit only a tiny fraction of their present level of pollutants. (Consumers, of course, can expect to bear much of the burden of the changes by having to endure higher car and gas prices.)

Even if results should not match promises, the IC engine is unlikely to get much immediate competition. The widely heralded electric car concept is still far from becoming a mass-produced reality. From an environmental standpoint, its biggest drawback is that it must be recharged from electric power outlets, which increases the load on notoriously pollutant-producing electric utility systems. (Nuclear energy will probably not overtake coal as the chief source of electrical power, according to current estimates, until the 1990s. There have been recent allegations that the environmental hazards from nuclear energy—thermal pollution, radiation—may be worse than from fossil fuels.) Steam cars also have many boosters. But Inventor William Lear, after investing $5.5 million of his own money to devise a workable steam system, announced in late 1969 that he had "thrown in the sponge" on steam and that there was only "one chance in 500,000" that a replacement for the IC engine could be found. Perhaps these problems will be solved or some radically new device, such as a fuel cell, will be perfected and accepted by Detroit. But it will likely be several decades before a radically new engine can be introduced into all new cars and most of the old IC engine autos—there are currently 100,000,000 on the road—retired. A recent *Fortune* study noted: "Even if alternative modes of propulsion do emerge, the sheer economic, political, and social inertia represented by all the institutions that have grown up to feed and care for conventional automobiles would seem to preclude a sudden shift." Detroit will be the last to try to do away with those institutions. Many oilmen feel that even before the use of petroleum for energy becomes obsolete most of the country's production will have been diverted to the manufacture of such products as chemicals and synthetic foods.

The fallacy, however, of linking oil shale's future with the estimated future supply of crude is that crude oil and shale oil, though perhaps more or less interchangeable at the refinery gate, are two different, distinct, separate products. One does not pump oil from shale deposits. One mines and retorts shale oil. Shale oil, according to the evidence, may well be a lot cheaper to produce. It may also have significant technological advantages. Shale oil for instance is very low in sulfur and thus is much less of a pollutant than most crude oil. And while research is still scanty, some shale engineers, such as Russell J. Cameron of Cameron Engineers, have recently come to feel that shale could become a major source of sulfur-free natural gas, which can be produced from shale quite easily. Shale gas, Cameron says, might be very competitive with liquefied natural gas imports, on which the United States, because of dwindling domestic gas reserves, is becoming increasingly dependent. To say we do not "need" shale oil, as oilmen assert, is a little like saying we do not need synthetic fibers like nylon and rayon because we already have plenty of cotton and wool, or that we don't need to have Chrysler manufacture automobiles because General Motors and Ford could easily take care of Chrysler's share of the car market. Dr. Orlo Childs of the Colorado School of Mines put the argument this way to the Senate Antitrust Subcommittee:

We do not "need" oil in large quantities from domestic sources. The crux of our problem is whether or not we "need" oil from oil shale as part of our domestic production. At this point we must distinguish "need" from "shortage." This year there is no shortage of conventional oil in this country. As indicated earlier in this presentation, a recognizable shortage may be a decade away.

What we really mean, or should mean, when we say there is a "need" for oil from oil shale is simply that there is such a "need" only if oil from oil shale can compete without subsidy in the open market with other energy sources. . . . (I)f oil from shale cannot compete—if it cannot carve out its own place in our energy market—we don't "need" it. But if it can compete —if it can bring our citizens better and cheaper petroleum products—then this is in the public interest and we "need" it.

At another point, Childs warned: "Let us not follow blindly those who say there is no 'need' from oil from oil shale unless and until they decide and determine how and when to let it enter the energy picture."

Equally fallacious is acceptance of the oil industry's statements on the potential profitability of shale oil production as being totally disinterested. John Kenneth Galbraith told the Senate Antitrust Subcommittee that: "The fact [the oil companies] are not out developing—this is an industry which is not normally immune to the profit motive—would indicate that it isn't very profitable." Galbraith rather naively ignores, for one thing, the very special economics of producing crude oil, which Chapter Two described in detail. While a shale oil operation's profitability might prove very attractive to the average businessman, the oil man, unless he is running short of crude, has numerous motivations to demur. Oil men do not mention in their evaluations of shale oil's profitability compared with crude that apples are being matched against oranges. The profitability of shale results from a conventional analysis of total capital investment, operating costs, selling price, and so forth. But the profitability of crude oil has built into it a large number of special advantages and disadvantages deriving from the oil industry's various economic subsidies.

This leads to one of the chief social advantages of shale oil over crude oil. Dr. Walter J. Mead of the University of California at Santa Barbara pointed out to the Senate Antitrust Subcommittee in March, 1969, that the abnormally high domestic price of crude coupled with the industry's tax subsidies acts to raise the after-tax profits from petroleum exploration and development to unnaturally high levels. But this then stimulates overinvestment, which eventually reduces the rate of return from investments in oil to the level of other investments. Thus the fact that the oil industry's net return on investment, *after all of its tax breaks,* is still just equal to other industries is actually proof of "resource misallocation," not, as the oil industry continually points out in newspaper advertisements, justification for its tax breaks. The result, Mead continued, is "submarginal uses of scarce capital" and "organized waste" by the industry. For instance, he cites bidding for Outer Continental Shelf leases off Lousiana in 1954 and 1955 which totaled $216 million in bonuses. If the domestic price of crude were to fall to the delivered price of imported oil at an East Coast port and if the oil industry bidders were subject to normal 48 percent income tax rates, Mead concluded that the after-tax returns on the investment even *excluding* the payment of the bonuses would have been only 4.5 percent. "These findings indicate," Mead said, "that if oil companies expected to earn more than 4.5 percent, and they do, after taxes on their oil development investment, they would have regarded the 1954–55 leases in total as having no present value." Estimates of shale oil's profitability generally assume sale of the shale oil at the inflated domestic crude price, and insofar as this is so, shale oil would carry an amount of resource misallocation. But in all other respects, specifically the tax subsidies, shale oil production theoretically entails none of Mead's "organized waste" and huge overinvestment disadvantages with commensurate social costs. And the closer shale oil comes to being able to compete with the delivered price of foreign oil, the nearer investment in shale oil will come to eliminating organized waste entirely.

If oil executives, further, agree that producing shale oil *without* subsidies, is "closely competitive" with producing crude oil, then producing shale oil would be considerably more profitable than crude, perhaps even spectacularly so if crude's subsidies were eliminated. This of course does not take into account numerous estimates by parties such as Tosco that shale oil could compete even on the basis of current inequities. Nor does it consider the possibly substantial economic benefit of the various sodium minerals intermixed with the shale. Shale oil producers, moreover, always have in their favor the downward trend of shale oil production costs as technology improves and the economies of scale are utilized, while crude oil producers face a long-term upward trend in their costs as existing reserves are depleted, and excess production capacity is cut back, forcing expenditure of greater and greater amounts of money on exploration and development of new reserves. A recent study by the Independent Petroleum Association of America concluded that a mere 25¢/barrel drop in crude prices would be sufficient virtually to eliminate exploration and drilling by independent producers, who account for half the nation's exploratory drilling. "Eventually, shale oil costs must decline relative to crude oil costs where the cost advantage of the former is controlling," Dr. Henry Steel of the University of Houston told the Senate Antitrust Subcommittee in 1967.

Many people with an interest in shale, such as Tosco, argue eloquently that oil shale should be given the same tax advantages as crude so that both could compete on a fair basis. While a case can be made for this point, it obscures another advantage that shale oil has over crude. Except to give it competitive parity, shale oil does not need tax advantages because it provides the solution automatically to all of the "national security" problems that oil men use to justify their subsidies. The location of the shale is well known, so there is no need for stimulating exploration. Since it is located in the United States, it cannot be cut off by Arab countries angry at the United States' policy toward Israel. Since it

is virtually inexhaustible, there is no need for special measures to ensure a protective reserve supply. "Oil men have to turn their national
security doxology around when they talk about shale," says a government oil economist. "They like to claim it's worth paying more for
crude because of the national security benefits. But when they discuss
oil shale which would have the same benefits, they say you shouldn't
have to pay the extra money it might cost to bring shale into commercial production."

Oil men might respond to these arguments by saying that however
overinvested the oil industry is, without its special benefits there would
not today be a sufficient crude supply. They would add that however
optimistic oil shale people are about shale's production potential, it
could not possibly supply all of the country's petroleum needs. Indeed it
will be lucky to supply even a third, and the oil industry will need
plenty of incentives to find and develop the other two-thirds. The answer to this contention is that if production from shale were to grow to
even a few million barrels daily, as the hypothetical sequence of events
outlined in Chapter Two demonstrated, this would not destroy the oil
industry but would force it into a series of reforms that would be bene-
ficial both to the nation and to itself. An April, 1965, *Fortune* article
entitled "U.S. Oil: A Giant Caught in Its Own Web," suggests that if the
$2 to $4 billion worth of waste built into the industry were eliminated
by junking the various "conservation" techniques and by affecting various other reforms, "Much of the inefficient domestic industry would
disappear, but the rest would survive, leaner and more efficient. The
very efficient producers would be vastly better off; operating at a higher
rate but at much lower prices, they could make more money than they
do now." The result would be substantial benefits to the consumer. The
cartel-like hold of the oil industry on petroleum production would be
broken. Both crude oil and shale oil could vigorously and profitably compete at some lower price level determined not by a maze of artificiality
but by capitalistic competition. And the lower the level of prices was,
the more consumers would benefit in lower prices for gasoline, fuel oil,
and hundreds of other petroleum products.

The final advantages of shale oil over crude oil are environmental.
Recent scientific attention has been focusing on increasing pollution of
the oceans by crude oil which, according to a *Wall Street Journal* report, "is nearing crisis proportions." Some of this pollution derives
from the fact that about 60 percent of the world's crude output is transported by sea. It is a common practice, for instance, for tankers to flush

out their cargo tanks with salt water. There have also been a growing number of ship accidents, such as the Torrey Canyon disaster in 1967, which are likely to continue to increase as imports to the United States grow and as more oil is transported in huge supertankers. Much oil pollution of the seas has been caused by offshore drilling. A Presidential panel on oil spills estimated that by 1980, 3,000–5,000 offshore wells will be drilled annually and that the country can expect a "major pollution incident" like the one in the Santa Barbara Channel "every year." These incidents do more than blacken beaches and kill a few sea birds. Some scientists feel oil pollution may gradually upset the ocean's entire ecology and preclude its use as a major food source. Already, some contend, cancer-causing agents present in petroleum are accumulating in sea life and may be in the process of being passed on to humans.

Nowhere is the environmental threat from crude oil as great as in Alaska. As one conservationist has written, "If oil is a uniquely devastating ecological enemy, Alaska is also a uniquely vulnerable victim." On its frozen tundra, for instance, debris does not easily decompose and scars do not easily heal: they remain for many years. The threat to the area's delicate environmental balances from thick pipelines and routine transport of oil by huge tankers through the Arctic icepack is massive. If the Prudhoe Bay finds lead to further exploitation of Alaska and other Arctic regions, some of the planet's last remaining sections of wilderness may be irretrievably damaged.

Production of shale oil in the Piceance Creek Basin will admittedly deface the landscape—waste shale will probably be deposited in canyons and then revegetated. But the long-range likelihood of serious environmental damage is minimal. For one thing, the operation will entail not the pumping of liquids but the more controllable mining of solids. The main potential pollution problems, such as contamination of the area's water supply, lie not in possibility of unavoidable accidents as in the case of crude production but merely whether or not sufficient antipollution safeguards are imposed on shale oil producers. The on-site presence of pollution-control agents such as nahcolite could substantially reduce the cost to producers of these safeguards. (The use of atomic *in situ* techniques would raise many grave environmental issues— mainly release of radiation—and may have to be avoided altogether. *In situ* fracturing by conventional explosives might prove feasible, however.)

The principal difference in environmental effects between the two materials is that crude production entails continuous movement: the drilling of new wells, exploration for new oil fields, construction of new pipelines. Progressive imposition on new regions of the globe, principally wilderness, is unceasing. On the other hand, once a shale oil production complex is established in the Piceance Basin, or perhaps Utah, it can remain stationary within a confined, controlled area, for a century or more. In short, for crude production, the environmental unknowns never end. For shale oil production, they can be swiftly dealt with and eliminated.

Whether all of the potential social benefits of a shale oil industry will ever come to pass will depend on just who the future shale oil producers turn out to be. In the absence of a specific federal shale development policy, the most likely chain of events, as distinguished from Chapter Two's "conceivable" chain of events, is this: The key to the beginning of shale oil production lies at the moment with Atlantic Richfield. Arco seems sincerely interested in aggressively confronting the uncertainties of oil shale, presumably on the theory that, at least to companies without large foreign reserves, oil from shale will sooner or later be a necessity and that the first man usually can get the best deal. Arco probably is consoled by the expectation that operation of the initial commercial ventures by oil companies will likely preclude the appearance of non-oil companies, and ensure the industry's long-range control of shale production.

The rest of the oil industry has resisted these ideas, feeling that it doesn't need shale yet and that the safest as well as the immediately least expensive course is just to ignore shale. Let shale be the problem of the next generation of top executives, which in the oil business is often only a few years away. But if Atlantic Richfield and the other Colony venturers really get a commercial shale plant going, it can be expected that a few of the other more aggressive major companies, such as Mobil, Shell, and Humble, might begin shale plants, too, to protect themselves competitively, if nothing else. Most likely they will license the Tosco process, for there does not appear to be any other process around nearly so advanced. Its technological superiority and its potential profitability seem fairly clear. Huntington Hartford will at last have lived up to his forebears. A quick breakthrough in *in situ* is possible, but widespread *in situ* retorting will not be possible until the government lands are opened.

Once production has begun, it is extremely doubtful that many non-oil companies will enter the field. The oil industry's stake in the presently available private land and water rights already provides it with a strong de facto control, and entrance by an outsider not possessing any of these things will become increasingly expensive. So far, with the exception of Cleveland Cliffs, not a single major corporation outside the oil industry has demonstrated even a remote interest in becoming a shale oil producer. Large mining companies have been nowhere in evidence. Kaiser Aluminum was the only major aluminum concern to make a thorough investigation of the possibilities of producing aluminum from dawsonite, but now it has dropped out. Many of the large coal companies have been acquired by oil companies. No natural gas company is known to have acquired shale land or conducted shale research (though the American Gas Association has made some rudimentary studies). The only large chemical company to show an interest in shale was Dow Chemical, which purchased a large block of shale land in the 1950s. But now Dow has agreed to sell the land to the Colony group. Even a firm like Koppers Company, which possesses just the capabilities and qualifications one would logically expect of an aspiring shale oil producer, says it is willing only to join a cooperative venture with an oil company. Even a large non-oil company interested in shale would likely be deterred by the manifold problems of competing with the $100 billion oil industry. Either it would have to assume the subservient role of today's independent crude producer, docilely dependent on the willingness of the integrated companies to buy its shale oil, or it would have to be prepared to bear the cost of establishing its own refineries and pipelines. Cleveland Cliffs' ambitions are admirable, but it is doubtful whether the company's financial resources will allow it to play anything but a fairly minor role. When I mentioned to Dayton Clewell, senior vice-president of Mobil Oil, that some people felt Tosco might be a competitive threat if their shale oil process were cheap enough, he scoffed at the idea. "Whom are they going to sell their oil to?" he asked. "Let's face it," says Russell Cameron of Cameron Engineers, "you're just stuck with the oil industry."

The effect of oil industry domination on an emerging shale oil industry is obvious. The incentive would be very strong for them to expand production (Louis Davis of Atlantic Richfield openly admits the speed with which Arco will bring on commercial production depends on the "need" for shale oil) and informally set prices in such a way as to avoid disruption of their crude business. Who will be able to prove them wrong when they say that the cost of producing shale oil just happens to be such that the selling price must be, coincidentally, the same as crude? Or that they have the resources to invest in a productive capacity of only so many barrels a day, which, coincidentally, happens to equal the domestic gap between crude supply and demand? In the end, the oil industry will have added oil shale as just one more component of its efforts to become a "total energy" industry; it already controls natural gas and coal and is making headway into nuclear power.

It is possible, though not likely, that a positive government leasing policy would alter these prospective events. Industry's response would of course depend on the terms.* Dr. Henry Steele suggested in an October, 1968, article in the *Natural Resources Journal* that the industry would want to be certain of what Charles F. Jones called the "running rules" before participating:

It is likely that a major factor impeding entrée into shale oil is the failure of government even to adumbrate any conscious policy for the phasing of shale oil into the national energy market. . . . One suspects that many potential entrants . . . are awaiting government assurance, in the form of explicit and detailed policies, that the advent of shale oil on the scene will not be allowed to depress crude oil prices. Such assurances would probably have a much greater impact on willingness to enter the industry than would even the announcement of a quite liberal leading policy.

As Steele pointed out, Stewart Udall seemed to have similar assurances in mind when he told the Senate Antitrust Subcommittee in 1967:

. . . if we handle the development of this resource wisely, if the Government maintains control, as it must—oil shale ultimately, when you look down the road, will be the dominant factor in this larger picture. The Federal Government, by controlling the rate of development of oil shale can in effect much more wisely determine what its energy policies should be.

*   *   *   *   *

We will decide that in terms of a national energy policy, that we want to bring in a certain amount of production, and we will put it out for leases, and on a competitive basis of some kind, perhaps. This is probably what we would do if we had the process perfected today.

*   *   *   *   *

* Detailed discussions of possible leasing proposals are: *Legal Study of Oil Shale on Public Lands,* a report for the Public Land Law Review Commission by the University of Denver College of Law in 1969; articles by Dr. Walter J. Mead of the University of California at Santa Barbara in *Natural Resources Journal* (April, 1967; October, 1968), *Colorado School of Mines Quarterly* (October, 1968) and *Proceedings of the Rocky Mountain Petroleum Economics Institute* (sometime in 1970); and articles by Dr. Henry Steele of the University of Houston and George Miron, former Associate Solicitor of the Interior Department, in the *Natural Resources Journal* (October, 1968).

[The Canadians] limited their production [of tar sands oil] in order to phase it into their whole energy economy, in order to not disrupt it. Now, this is the type of problem that we are going to have down the road when we start talking about how oil shale development would fit into the total national energy problem . . . [The Canadians] can control it, just as we would be able to control oil shale—so as to produce the minimum disruptive effect on their energy economy.

Even if the government were to make such promises, the oil industry might be very skeptical and unwilling to relinquish that much power over its destiny. But if at the time such promises were made several large oil companies had begun producing shale oil commercially, it is very possible that the industry would be receptive to a variety of leasing proposals. The more certain of the future the industry is, the more responsive to leasing plans it will be. Many oil executives have recommended that open competitive bidding on tracts of shale land be conducted at regular intervals, much as offshore land is auctioned off. Industry reaction to such a plan would probably indeed be very favorable, for it would simply be a means of gaining control over more shale land which it could then develop if and when it saw fit. If the government injected restrictions to require development or high rental payments to penalize nonproductive holding of the land, industry might acquiesce to a point, but if the terms became too tough, it would boycott the bidding as it did in December, 1968.

Guesses on what future leasing proposals might be like must reckon with the Teapot Dome Syndrome. As Udall said, "If Hickel drafts anything that will attract anyone, a lot of people are going to call it a giveaway. And they may be right." Notice, for instance, the criticism from Grabber Fighters and others he received for his 1967 and 1968 proposals which had only the faintest attraction to the oil industry. It is probable that the willingness of the public and the vocal crusaders for the public interest to accept any leasing proposals attractive to the oil industry is likely to decline. Despite their apparent anxiousness to use this great wealth for grand national purposes, they appear almost more willing to have it stay in the ground rather than risk the chance it might provide a windfall for private developers. But if the oil industry were to begin substantial shale oil production on the private lands, pressure for a workable leasing program would become irresistible. If one were devised, it would be exceedingly difficult to avoid leasing the public land to anybody but the existing shale oil producers.

There are only two ways out of the foregoing prospects. One is the creation of a special governmental or quasi-governmental corporation modeled after TVA or the Communications Satellite Corporation. The most notable advocate of this idea has been Dr. Morris E. Garnsey, professor of economics at the University of Colorado. He told the Senate Antitrust Subcommittee in 1967 that "the oil industry in the United States is so highly monopolized that we cannot trust the development of this resource to that industry." The only way to prevent monopoly, he said, was "to introduce a new competitive force into the industry in the form of a government authority for the operation, exploitation, and development of oil shale and related minerals." Garnsey said he favored the ComSat idea because it "comes closer to joining public and private interests in a single business unit that any other commercial organization ever created." Besides prevention of monopoly, he said the concept was also justified by "the vastness of the resources involved and the consequent scope of the endeavor necessary to develop them; the uncertainty and novelty of the technology; the substantial and distinct public interest; and the necessity for continuing cooperation between government and industry in the development of the resources."

The political feasibility of the idea, however, seems very remote, for it would mean government entrance into a business clearly within the technological capability of private industry and in direct competition with private industry. Whether the government, with its huge vested interest in other energy areas such as atomic power would be much more responsive to the real public interest in such a corporation than any wholly private party is not altogether certain.

The only remaining possibility is an extensive government research program to create commercially-applicable technology and patents, which could then be put in the public domain. A similar idea was one of the options mentioned in a September, 1969, Bureau of Mines report to the Cabinet Task Force on Oil Import Controls, which evaluated the cost to the government to subsidize the development of a strategic oil reserve if import controls were eliminated and the domestic price of crude were allowed to drop to $2 a barrel. Based on figures in Interior's May, 1968, report on oil shale, the Bureau estimated that to secure an eventual production of one million barrels/day from oil shale, the government would have to build three prototype oil shale plants involving different retorting processes, each capable of producing 35,000 barrels/day of semi-refined shale oil. Capital costs would be $415 million. The plants would begin operations, if design began immediately, between 1973 and 1975, and then be turned over to private industry.

Meanwhile by 1974, the report said, the technology and cost estimates hopefully would have been proved to the point that private industry would start investing their own money in similar plants. Initially, with the domestic price at $2, the government would have to subsidize the industry at a guaranteed rate of 98¢ a barrel in order to allow shale oil producers to earn a 12 percent return on their average investment. If improved technology developed by 1978, the subsidy would fall to 12¢ a barrel, and by the 1980s, no subsidy at all would probably be required.*

A less ambitious program was proposed by the Bureau of Mines, the Geological Survey, and the Bureau of Land Management in February, 1967. It recommended that $101 million be spent over a ten-year period on an effort whose primary emphasis would be "the development of new or improved technology." Senator William Proxmire of Wisconsin and five other senators introduced the Oil Shale and Multiple Minerals Development Act of 1967 which outlined a similarly broad federal research and investigation program.

An aggressive government research program would have two main advantages. It would produce detailed data on production costs, which would allow the government to design an equitable leasing program that would be relatively immune from giveaway charges. And if the technology were made widely available, there would be at least the outside chance that non-oil interests would be lured into shale production. The only way to ensure that non-oil interests then would be able to dominate the industry would be to offer them temporary tax advantages while using the antitrust laws to bar entry by large integrated oil companies.

---

* Of all the alternatives evaluated by the Bureau, including subsidizing greater oil exploration, maintenance of the existing system, with and without prorationing and the storage of oil in steel tanks, oil shale was the least expensive, except for the remote possibility that enough oil could be stored in underground salt domes. Oil shale was also less than 40 percent as expensive as coal, which has often been thought of as a synthetic fuel competitor with shale oil, though most experts feel it is much more costly to produce, mainly due to the large amount of hydrogen required. The oil industry has never been friendly toward coal and they forced closure of a government coal research plant around the same time Anvil Points was closed. But coal poses less of a competitive threat since most coal lands are scattered and in private hands, and many large coal companies are now owned by oil companies. Some oil companies are therefore engaged in modest research to produce synthetic fuel from coal, and, if some form of synthetic fuel were unavoidable, would prefer it be derived from coal rather than shale.

Such a program would of course be bitterly opposed by the oil industry as a gross and heinous violation of the American tradition of free enterprise as well as an unfair penalizing of those private concerns who have conducted private research and who, by this time, may already be producing shale oil. Though "few industries rely so heavily on Government favors," Economist Milton Friedman has said, "few industries sing the praises of free enterprise more loudly than the oil industry." Opponents of the oil industry would have the job of justifying huge federal investment in an unknown black rock in Colorado when so many other pressing needs exist and when private interest has shown itself, as oil men will point out, "ready and able" to do the job by themselves. Of course the oil industry is still receiving tax subsidies amounting to at least $3 billion a year. And President Nixon felt no qualms recently about announcing a willingness to spend $3.8 billion over 10 years to revitalize the U.S. Merchant Marine. But both the oil industry and the Merchant Marine have a somewhat larger and more active constituency than does the shale oil industry.

This book ends, then, on a mixed note of optimism and pessimism. It does seem clear that during the present decade the Elusive Bonanza will at last begin yielding up some of its immense riches. Rifle, Grand Valley, DeBeque will boom and perhaps even Rulison, the ghost town, will burst back to life. "A beehive of activity," Louis F. Davis of Atlantic Richfield predicts.

Much less clear is whether oil shale's great potential benefit to our society will be fully realized, at least in our time. "Nothing contributes more to rising productivity and living standards over the short run and over the long run than low-cost energy," said a 1965 *Fortune* study of the petroleum industry. Unfortunately, it does not now appear politically possible to prevent shale oil production from becoming just one more producing division for a handful of giant oil companies, who can be expected to so subdue any possible Schumpeterian "gale" that only a few languorous zephyrs will remain. "Monopolization of this enormous asset," Senator Philip Hart has said, "would be a national tragedy." But not enough people yet care for anything else to happen.

All of the old oil shalers will probably have died off by the time the oil industry starts digging away in earnest at the black rock. But it probably won't be the sort of thing they would have wanted to stay around for anyway.

Appendix

The reason for this Appendix, as the Acknowledgments section noted, is the public controversy that developed over certain episodes of my investigation of oil shale while I was business editor of *Life* magazine. Rather than permit to stand a few somewhat less than informed published accounts which may influence how some people judge my assessments of oil shale in this book, I have here described the facts as I understand them. I have also given my interpretation of them, together with the generally contrary interpretations that have been offered by some of the other people who were involved.

Life magazine had done a story on oil shale in the early 1950s which described the work at Anvil Points, but my first knowledge of the subject came from material sent over by public relations representatives of The Oil Shale Corporation in 1966. Their pitch was that the "baffled men" were about ready to tame the "rugged mountains." I did not become interested enough to pursue the matter until I read the *Ramparts* story which described J. R. Freeman's thesis of the Giveaway Conspiracy. With its mysterious threats and attempted murders the story sounded only marginally credible, but just the possibility that there might be a really big scandal caused me to fly out to see Freeman in May, 1967. He pointed out to me the gunshot holes in the back of his truck and showed me his confidential files, but several weeks of interviewing and research into the legal background of the pre-1920 oil shale claims confirmed my initial skepticism. No land, I found out, had been "given away" since 1960, and the rulings in the *Freeman* v. *Summers* and *Virginia-Colorado* decisions showed that Freeman's assertion that all the pre-1920 claims were "phony" was a matter of legal interpretation, not indisputable fact. Finally, no one had yet proved the existence of an actual bribe or payoff.

I was about ready to give up on the story when I came across John Kenneth Galbraith's statement in the report of the Oil Shale Advisory Board that "the incentive to control oil-bearing acreage is thus, for the time being, much greater than the incentive to produce from it." If this were true, I thought, then something was seriously wrong with the "baffled men against rugged mountains" theory. I made further investigations, which led to my current thesis on the oil industry's attitude toward shale.

I explained my findings to some of the senior *Life* editors in September, 1967, and in October, I turned in the first 24-page version of my story. I detailed Freeman's Giveaway Conspiracy and the Fred March episode as well as my skepticism about Freeman's theories, but the chief emphasis was on the reasons why oil shale had not been developed, specifically "the pervading fear of the impact that oil shale development could have on American business." Then I added:

In order to maintain the profitability of the oil industry, the price of U.S. crude is kept artificially high by strict government-imposed import and production quotas. A flood of cheap oil from shale might drastically lower consumer gasoline and fuel oil prices, vastly improve the balance of payments and avoid this country's dependency on the whims of Middle East sheiks. But it could also introduce the ravages of the free market to the oil industry, drive down the price of crude, make useless billions of dollars worth of crude reserves and, conceivably, even drive much of the powerful $50 billion oil industry out of business, . . . [This possibility] has . . . prompted most of the large oil companies to spend millions of dollars obtaining control over most of the privately owned shale land and to constrain or stifle potentially productive research efforts. Though many oil companies own more than 10,000 acres of shale land containing billions of barrels of oil, they have almost without exception been unwilling, as long as ample supplies of crude oil were available, to gamble the necessary money to develop a commercial facility.

Looking back on that first version today with the hindsight and increased knowledge of 18 months more research, I think it somewhat overstated and oversimplified the facts, though in a sense this is almost inevitable when one tries to squeeze something as complex as oil shale into 5,000 words. Particularly, I think it gives the erroneous impression that the oil industry's attitude toward land acquisition and research is part of some well-organized master plan when in fact, as is so often the case in the oil industry, the major companies merely have acted individually in a fairly uniform way to protect and enhance their own best interests. The result, though, remains the same. Despite these quibbles, the essence of the viewpoint expressed in this first version is the same one I hold today.

The reaction of the *Life* editors to the piece was generally enthusiastic—Articles Editor David Maness said he thought it was a "blockbuster"—but everyone had recommendations on how the piece might be better organized or otherwise improved. Such recommendations are not unusual—*Life* is known for its meticulous text-editing, and few long articles run without at least one fairly major rewrite. I spent hours with Maness and other editors working to clarify points and make structural changes. After three revises had been submitted without a significant change in emphasis I detected some slight uneasiness in Maness and Assistant Managing Editor Robert Ajemian, my immediate boss. They had no quarrels with that section of the story which discussed the Giveaway Conspiracy. But they were worried over my more controversial allegations against the oil industry. They suggested that perhaps the tone of the story should be shifted. Instead of me, the author, pointing a finger at the oil industry, it might be better to have me, the author, as a disinterested reporter, describe both sides of the controversy.

Their concern, as I interpreted it, reflected genuine misgivings about the validity of my point of view, which of course was not nearly so comprehensively developed and documented as it is in this book. But it was also based on the fact that, before reading my piece, they had never before heard about the oil shale controversy. To my knowledge, the only previous press accounts to touch upon the oil industry's lack of interest in shale development over the past ten years (there was a flurry of comment when Anvil Points was closed) were "Shale Oil—The Cartel's Ace in the Hole" by Roscoe Fleming in the March 15, 1965, issue of *The Nation* and "Bonanza in Colorado—Who Gets It?" by Julius Duscha in the March, 1966, issue of *The Atlantic Monthly*. Neither article had emphasized or developed the point, however. There had been nothing in *Time* or *Newsweek,* and nothing in *The New York Times* except for a few routine stories on Udall's leasing proposals by William M. Blair. *Life* has no hesitancy about going after the Mafia or even Abe Fortas because the *point of view* is accepted: organized crime has great power, public officials act illicitly. But before running a controversial story with as alien and novel a point of view as my article on oil shale, the magazine appreciates the comfort of prior advocacy somewhere else. For instance, when Seymour Hersh, a former Associated Press reporter, became the first writer to uncover the details of the My Lai massacre in Vietnam and tried to interest *Life* in the story in November, 1969, he was turned down. But after the story developed into a *cause célèbre* and received voluminous coverage by all the media, *Life* rushed out to pay a sum reported to be near $25,000 to Ronald L. Haeberle for his exclusive pictures of the incident and ran a 10-page lead story.

I was not terribly happy about the suggested change in tone, but, as a veteran of five years' worth of group journalism, I acquiesced since the real thrust of my story would remain pretty well unchanged. Version #5, which was approved by David Maness, stated that "there have been dark hints of skullduggerous power plays by large oil companies and of scandalous misconduct by high Washington officials." The view I had stated as my own in the first version became the contentions of a "growing, very vocal group of critics" which I rather artificially lumped together. Then I presented rebuttals by the oil industry. By December, a version had been approved by everyone, including then Managing Editor George P. Hunt, and tentative plans had been made to run it in an issue right before Christmas. However Hunt then told me he had decided to delay it until January because it might not fit in too well with the holiday tone of the December issues.

In January, *Life*'s editor, Edward K. Thompson, was replaced by Thomas Griffith, a Time Inc. veteran who had been an editor with *Time* and senior staff editor for all Time Inc. magazines. (Technically, the editor is superior to the managing editor, though the managing editor retains fairly autonomous control over the day-to-day operations.) During that month, Griffith, Ajemian, Maness, and I had some long conferences during which Griffith raised a number of questions, many of them pertaining to the piece's structure and the way the arguments were developed. He was also unsure about my description of what might happen if a sizable volume of shale oil should enter the crude market and my discussion of the possible effect on the price structure of crude. On January 18, he wrote me a memo which read in part:

Let me try to put down what I think the story is, and how it might go.

1. An opening, pretty much as now, down to the middle of I-2: which makes a vivid scene setter. Yet though long known, these riches have been virtually untouched (I-3).

2. Why all this isn't being used, why it isn't bringing lower gasoline and fuel oil prices at every corner pump is the subject of a very heated controversy these days. Full of dark hints at the mysterious power of the big oil industry, suspicions of skullduggery in the hills and shenanigans in Washington.

3. The big oil companies once ignored and pooh-poohed shale oil, then belatedly tried to buy up claims. But they still say it's too costly to extract: quote Humble. They are spending very little on research; and the attempts of two oil companies to do so were called off in circumstances that many industry critics—not to mention property owners, politicians, investors, and a lot of other people in the Rockies—find highly suspicious. . . .

This is one way the piece might be organized: not the only way, or one you need be bound by if difficulties develop in the telling. I'm also making some specific marks on the hard copy. In general, I think there's a good story here, and some good reporting, which will be strengthened once we separate out the anonymous and unspecific "could happens" from what, in our best judgment, seems likely to happen, so that the reader has a clearer understanding of what's going on in a very complex situation. Far from wanting no judgments in the piece, I think the reader needs guidance on what weight to give this or that necessarily anonymous remark.

In version #7 I incorporated Griffith's suggestions. In the meantime, however, a memorandum had been received from Gilbert Burck, associate editor of *Fortune* and author of two long stories on the oil industry in 1965, who had been asked for his comments on my shale story. Burck said that he had "mounting doubts" about the theory that the oil industry was opposed to the idea of oil shale development: "I fear there is something wrong about the basic approach. . . . I can see the possibility of shale oil developing into something great, and just possibly providing competition with the Texas-controlled companies. But to say they don't want to spend money developing cheap shale oil because they have too much crude oil already just isn't so."

I am certain this criticism weighed heavily on the minds of *Life* editors, for around Time Inc. the business and financial expertise possessed by *Fortune* editors is considered to be preeminent. Burck had been writing for *Fortune* for 28 years and had been a member of the Board of Editors since 1943. By contrast, I had been *Life*'s business editor (the first the magazine had ever had) for only two years. I had no special qualifications or "credentials," to use a favorite Time Inc. term, such as a degree from the London School of Economics. I had majored in politics at Princeton, had no advanced degrees, and had been writing about business for only three years. Prior to that I had been a reporter in

Life's entertainment department covering television and popular music.

Additional conferences ensued, and Griffith's doubts, like Burck's, seemed to be mounting. He seemed particularly concerned about my detailing of the relationship between Gulf Oil and the closing of Union Oil's oil shale plant in 1958. He contended the evidence was circumstantial, and that I had not "proved" that Gulf had caused Union to close the plant. I said I felt that the evidence still was substantial, and that *Life* often, for example, states flatly that people are "hoodlums" or "racketeers" for the Mafia on the basis of such circumstantial evidence as records of phone calls to Las Vegas and association with known Mafia "chieftains." I added that despite this, the piece was now structured around the device of charges and answers, and it seemed perfectly good journalism to report what Tell Ertl and others claimed had happened and then quote denials from Union and Gulf. Griffith still felt that somehow we should be able to prove Ertl's case; otherwise, he said, the whole theory of the industry's hostility to shale research might be suspect.

Around the end of January, I first heard that the *Life* publisher's office and the advertising department were quite interested in my story. During my time as business editor, executives on "the business side" had usually read my stories before they were published and had occasionally offered suggestions. George Hunt, however, had frequently told me the editorial side should never feel bound to accept suggestions from the business side if we did not feel they were editorially justified. This policy was scrupulously followed in the dozens of business stories my department produced prior to oil shale. Once my assistant, Nancy Belliveau, wrote a story on a new advertising campaign for American Motors devised by Mary Wells of Wells, Rich, Green, at the time Madison Avenue's most talked about new ad agency. The opening picture in the layout showed a scene from a planned TV commercial in which a group of workmen were demolishing a Ford Mustang with sledgehammers. They then reassembled the parts into a Javelin, AMC's new sporty car. The text quoted Ford Vice-president Lee Iacocca as saying, "She'll lose. Anytime people try to play dirty, they will lose, even though they might gain a temporary advantage. The public is too smart for that type of approach." The advertising people were not happy that the story in effect ballyhooed an advertising campaign that was largely devoted to television, but they were specifically unhappy about the Iacocca quote which they said was not really representative of what Detroit thought about

Mary Wells and should therefore be deleted. I argued it should stay in, and it did.

The initial suggestion from members of the ad department on the shale story was that the oil industry's response to the contentions of the critics should be given additional space. When these views were relayed to me, I obligingly added a few paragraphs. But apparently this was not sufficient. A number of meetings were held between the business side and the editorial side, to which I was not invited. I was told the business side did not want the story to run, but that Griffith had made a truly memorable defense of editorial integrity in general and my story in particular. George Hunt informed me that *Life* publisher Jerome S. Hardy (now departed) had compiled an estimate that if the story were published, *Life* would lose between $5 million and $20 million worth of advertising, presumably from angry oil companies, but that I should not be concerned. Another senior editor told me that Shell Oil had decided to delay signing a contract for a series of two-page color ads scheduled to begin in August until, the Shell executive told a *Life* ad salesman, "we get a chance to see how the story turns out." Nevertheless, it appeared the business side had retreated, for on February 1, 1968, the story "closed" for publication in the issue dated March 1. George Hunt was out of the country at the time but the story could never have closed without his, or Griffith's approval. Covering four full pages and three half pages, it would have been that issue's major text piece. (When *Life* closes a story, this signals the end of the editorial process. By that time all of the writing, editing, and layout have been completed and approved by the top editors and the story has been sent to Chicago where it is made ready for the presses. The only further connection with the story by the New York office is proofreading of the "page proofs" after the story has already been set into type. The rare exceptions are usually caused by fast-breaking news developments which require a few additional facts to be inserted in a previously closed story or even require the old story to be replaced entirely by coverage of the new development. Interruption of the post-closing process for other reasons almost never occurs.)

On Thursday, February 15, after page proofs as well as proofs of the color engravings for color pictures on the first three pages of the story had been received in New York, George Hunt, who had returned to his office the previous week, called me into his office to tell me that "a number of points" had come up, and that he had decided to take the story out of the March 1 issue and put it in the March 15 issue, which would close the following Thursday. Hunt did not seem to me to be too clear on just what the "points" were. He mentioned the Union-Gulf episode which I thought had been approved by everyone. (As version had piled upon version, I had begun to feel in fact that actual improvement and clarification had given way to a certain arbitrary shuffling of thoughts and paragraphs from one part of the story to another. At one point, I was asked to cut back on stylistic and colorful writing so as not to "prejudice" the various points of views. I went through the copy deleting dozens of adjectives and adverbs and toning down verbs. But the result was criticized for being "too dry and too dull" and I was told to put a little life back into the piece. I had the feeling that each version was being given to newly formed groups of editors, whose suggestions did not necessarily have any relevance to improvements made upon the suggestions of previous groups of editors.)

Hunt then said that the advertising people had recommended that the story be sent to the major oil companies for their comments. Imagining the endless delay that would obtain if we waited for a dozen or so companies and their attorneys to peruse the piece and prepare doubtless lengthy replies, which we would then have to figure how to incorporate into the story, I said I felt the oil industry's position had already been well presented having been based on numerous interviews I had had with oil executives as well as public comments in speeches and testimony before Congressional committees. Finally Hunt and I agreed I should try to "sharpen and clarify" the differences between the critics and the oil industry. I suggested we might add subheads such as "The Oil Industry Position." He replied he thought that was a very good idea.

The following Tuesday, I submitted Version #9. It read in part:

On one side of the controversy in the powerful $50 billion oil industry, which has long maintained that it isn't worth the expense and effort to produce oil from the shale. On the other side is a disorganized but increasingly vocal group of critics—staff members of various congressional committees, frustrated owners of shale land, investors, professors of economics—who angrily accuse oil men of protecting their crude business by acting to deny the country the obvious and substantial benefits of shale development. Even more villainous, in the view of the critics, is the Interior Department, which administers that portion of the shale (80% of the total acreage) owned by the federal government. Interior, it is alleged, has neglected the public interest and instead favored the obstructionist tactics of the oil industry.

At stake in the swirl of charges and countercharges is nothing less than this basic issue: Will oil from shale replace crude oil as the nation's principal source of energy?

THE OIL INDUSTRY POSITION

The oil in oil shale, the industry argument goes, is somewhat analogous to the estimated $7 trillion worth of gold in sea water: an alluring treasure until you try to get your hands on it. "No practical oil man will be misled by what may seem to be a mammoth supply of oil in the shale, just ready for the taking," said C. E. Reistle, Jr., former board chairman of Humble Oil in a speech. "There is a big difference between this 'on the books' oil and oil in the pipeline."

Simply to come up with a production process that would be competitive with crude oil, Reistle said, would require 8 to 10 years of work and the commitment of "large sums" of research money. Once the process was developed, many more millions of dollars would have to be spent on commercial plants and equipment. Such an extensive search for probably very modest profits has seemed quixotic when the oil industry already has nearly $200 billion worth of crude reserves just waiting to be pumped from the ground.

Furthermore, petroleum production is a drilling and pumping operation in which the industry has invested billions in equipment and technology. Producing shale oil involves the alien, to oil men, disciplines of mining and "retorting," whereby tons of rock must be crushed and heated to yield a few gallons of shale oil. And this shale oil would not qualify as crude does for the $27\frac{1}{2}$ % depletion allowance—basic to oil industry profits—which allows crude producers to deduct from earnings $27\frac{1}{2}$ % of their gross income.

There is always the possibility, though, that someday somebody might invent an economical shale oil process or that world supplies of crude could be used up. The strategy of the oil industry toward shale over the past 40 years thus has been to buy up or obtain control of most of the privately-owned shale land and then wait for the most opportune time to develop it.

THE CRITICS' POSITION

The critics are outraged that such a rich national asset should continue to lie dormant, especially since a number of government geologists, shale engineers, and economists have predicted that without much difficulty shale can be made to yield oil at costs below that of domestic crude.

Numerous benefits follow if this prediction is correct: lower gasoline and fuel oil prices, less national dependence on limited domestic reserves and those in often unstable Middle East countries, a huge new source of federal revenues. And the elimination of oil imports would solve the country's balance of payments and gold outflow problems.

The real motivation for the oil industry's strategy, this argument goes, has been its fear that by developing its shale, it would be inviting possibly dangerous competition. The federal government has not often been known as a zealous foe of the oil industry. But if shale production proved commercially very attractive, the government might open up federal land to companies not associated with oil interests.

At hearings last year before Michigan Senator Philip Hart's Antitrust and Monopoly Subcommittee, University of Houston economist Dr. Henry Steele testified that the "irresistible pressure" of large quantities of low-cost shale oil could weaken the complex system of federal and state regulations which supports oil industry profits and conserves domestic crude reserves by keeping crude prices artificially high. A respected energy expert from a private Washington research organization adds that such a development could "make Texas and Oklahoma look like another Appalachia."

University of California economist Walter Mead told the Hart Committee that even if oil companies were provided with a sure-fire way to produce cheap shale oil, most would resist rapid shale development because this could threaten high crude prices. The value of their reserves, said Mead, "depends on the price of oil. Any firm with large reserves is not terribly interested in seeing the price go down."

As support of these assertions, the critics point out that despite the industry's expenditures of millions of dollars to buy up shale land, most companies—especially such crude-rich corporations as Gulf who have the most to lose from a flood of cheap shale oil—have spent only minuscule amounts of money on research to see if commercial development were feasible. In fact, the critics go on, some members of the industry have throttled or gained control of several research efforts when they began to look promising.

As an example of the sort of thing they mean, shale critics point to an episode involving Union Oil of California and Gulf Oil. . . .

When I handed this version to Hunt, I said, "I'm really beginning to get worried about this now, George. I just don't think you plan to run the story at all."

"Look, don't worry, Chris," he replied. "I'm telling you: this story will close this week and that's *final.*"

For two days I heard nothing. Then Thursday evening about 6 P.M., I was called into Hunt's office. Present were Hunt, Griffith, and Ajemian. "We're terribly sorry to have to tell you this," Hunt said as I sat down, "but we're just not going to run the shale story."

"You're not going to run it at all?" I asked incredulously.

"That's right," Hunt replied. "The story is dead."

Griffith explained some of the reasons for the decision. No advertising pressure was involved, he said. It was just that he still had too many doubts. He went over many points in the story which we had previously discussed in much detail, especially the Union-Gulf episode. Can we really believe this man? he asked. How can we be sure of this incident? Is this fact really reliable?

"So what you're really saying," I declared, "is that unless I can somehow come up with a tape recording of the president of Gulf telling the president of Union to shut down his shale plant because he's worried about competition from shale oil, then we can't run the story. Is that it?"

Griffith just sat there and looked at me and said nothing. Then he mentioned he was late for an appointment and left.

I turned to Hunt. "How could this happen?" I asked. "You *promised* me the story would definitely close this week. How can you all of a sudden say that after going over the story for months and finally approving it suddenly you developed some doubts about things I thought were cleared up long ago and the doubts were so big you had nothing else to do but kill the story? How can you say that?"

"I know that, I know that," Hunt said very quickly. "You don't have to rub it in. I'm embarrassed enough about this as it is." He indicated that while the managing editor has wide control over the daily running of the magazine, the editor retains the power to exercise a veto.

To this day, I am not really sure why the story was killed. Numerous explanations were offered when the episode later became a matter of public discussion. A *Wall Street Journal* article which said that I thought pressure from advertisers may have been responsible quoted *Life* publisher Jerome S. Hardy as replying that, "Chris just couldn't be wronger." George Hunt said, "It didn't turn out to be the definitive piece on oil shale that we wanted to do. . . . (T)he piece required a kind of sophistication that just doesn't come through." Hunt admitted that this kind of story could effect advertising revenue, but "that had nothing to do with our killing the story." Thomas Griffith was quoted in *Advertising Age* as saying, "If we are satisfied with the story and feel we have it straight, and have confidence in our story, we print it. That just wasn't the case with this one."

J. R. Freeman and some of the other Grabber Fighters made the issue the subject of numerous speeches and newspaper articles which claimed that powerful Grabber forces had acted to squelch national exposure of the Giveaway Conspiracy. According to an editorial entitled

"Life Magazine Knuckles Under" in the Fort Collins, Colorado, *Northern Star:*

. . . (A)t some time prior to the March 18 date, someone in a very influential position in Shell Oil paid a visit to Secretary of the Interior Stewart Udall, and Udall's chief Interior Department solicitor, Frank Barry. Speaking for all major oil companies, the Shell representative quickly solicited the cooperation of the Secretary of the Interior. And someone at *Life* magazine killed the story. . . . So, the American public still has not had an opportunity to read the full story of the biggest land swindle perpetrated against the American public in thirty years.

Freeman also gave copies of page proofs of the killed story he obtained from Jerry Landauer, who had written the *Wall Street Journal* piece and who had obtained them in turn from me, to columnist Jack Anderson. "This column has obtained page proofs," Anderson wrote, "of an explosive oil shale expose which *Life* magazine mysteriously suppressed earlier this year after spending a year digging up the facts. . . . It is reported that oil company advertising, especially Shell's, would have been withdrawn had *Life* published its oil shale articles."

More recently, J. R. Freeman wrote a story on the subject in the Summer 1969 issue of the *Columbia Journalism Review.* Despite his attempts to "place" me in the hands of experts and help me to "speak the language" on oil shale (I had *Life* pay him $400 for research help he provided me) I had "not reviewed completely" the files in his possession and my story was therefore, he said, "a mere surface approach to the oil shale and related mineral resource scandals." Still:

There is at least adequate basis to speculate that oil interests influenced *Life* to scrap their oil shale story. Certainly one thing is clear. The then Secretary of the Interior, Stewart Udall, did not want to have the oil shale scandal given wide publicity. . . .

Freeman, of course, sees the demise of my story as just one more piece of evidence of the power of the Giveaway Conspiracy. This seems doubtful, if for no other reason than that I never detected the slightest sensitivity among *Life* editors about that part of the story dealing with the giveaways and Fred March. I am certain I could without correction have virtually called Udall a crook and March one of the great American heroes. Further, my criticism of the Giveaway Conspiracy would probably, if anything, have aided the Grabbers.

Assigning precise responsibilities and motives for killing the story is still very difficult because of the rather disorganized, even haphazard chronology of events: the story's initial closing, then its cancellation, then its rescheduling, then its final cancellation. This was due, apparently, to the conflicting decisions being generated from the long series of meetings of the top editors at *Life* and Time Inc. and of representatives from the business side.

While I can understand how Griffith and some of the others still had doubts about whether the oil industry was really opposed to oil shale development, I cannot understand why this should have been a factor in the decision not to run the story. To repeat, in the final drafts of the story I, the author, was not accusing anyone of anything, nor was I taking a position on the oil industry's motives. I was simply reporting the legitimate existence of a controversy. This meant that the ultimate burden of proof lay not with me but with the parties to the controversy.

The only explanation I can offer is that the editors simply did not feel that oil shale was an important enough story to risk the loss of all that advertising, particularly in view of the magazine's declining financial health in recent years. Actually, I find it very hard to accept the $5 to $20 million figure produced by the publisher's office. Shell and some other oil companies might have mentioned that they were not excited about the story, but I think the publisher's estimate was mostly just a chimera offered up to scare the editorial side. It is certainly a lot easier to sell ads if the magazine confines its exposés to the Mafia, Abe Fortas, and other conventional areas of crime and corruption instead of discussing criticism of its major advertisers. In a reply to Freeman's article in the fall 1969 *Columbia Journalism Review,* I stated, "The article's cancellation was the result of self-censorship. Though less sensational than collapse under pressure from powerful conspiracies it is, in my opinion, far more odious."

My personal reaction to that Thursday meeting in Hunt's office was a mixture of anger and disappointment, and I resolved that I would not simply put the story away in a file drawer. After some thought, for I knew that it would probably cost me my job, I went to see Willie Morris, editor-in-chief of *Harper's* magazine, who had once edited *The Texas Observer,* an iconoclastic weekly that has never hesitated to be critical of the oil industry, and who I felt might be interested in publishing my story. I explained the entire situation, told him I was coming to *Harper's* without the permission or even knowledge of anyone at *Life* and left him the final version of the *Life* story to read. A few weeks later, Morris told me he wanted to run the story but that it needed some rewriting. The chief criticism was expressed in a memo written by John Fischer, a contributing editor:

This ms. sounds confused primarily because Welles hasn't made up his mind what position he wants to take—what ought to be done and who the real villains are, if any. Once he gets a firm point of view, he should then be able to organize his material coherently around it.

Harper's accepted the rewrite, paid me $500, and closed the story in the August issue. The story thus came full circle, with its point of view back to the first version I had turned in five months and nine rewrites before.

The first *Life* found out about my negotiations with *Harper's* was the third of July, about three weeks before the article was due to be published. I had planned to wait a week or so before telling anyone just to be sure that nothing could be done to cancel the story. But since Robert Ajemian was about to leave on a month's vacation, I went into his office about 5 P.M. and announced, "By the way Bob, I thought I better let you know that the oil shale story is going to appear in next month's *Harper's*." He appeared aghast.

Within minutes, a conference was convened between me, Ajemian, Assistant Managing Editor Philip Kunhardt, and Jack Dowd, Time Inc.'s chief legal counsel. Kunhardt ordered me to call *Harper's* and have the story removed. I refused. He said the penalties for selling the story on the outside could be "very, very serious." I said nothing. He wanted to know how I could have deliberately sold material belonging to *Life* to another magazine without obtaining permission. I responded that if I had asked permission *Life* would have refused. He asked me why I didn't at least tell somebody what I was doing instead of keeping it a secret. I said if *Life* had known of my plans they might have taken legal action to stop the article from running in *Harper's*. What I had done, I continued, was to commit a deliberate act of corporate disobedience for what I felt were higher ends. The others in the room said they felt my act had been one of disloyalty, not to mention outright dishonesty.

Kunhardt and Dowd then discussed the possibilities of legally enjoining the issue of *Harper's* and preventing its distribution. Dowd said this would be difficult and might cost a lot of money. I was excused while everyone went over to George Hunt's office. On the way, Kunhardt telephoned Publisher Jerome Hardy whose office was on the floor above and said, "Jerry, you've got to come down here right away. Something terrible has happened. The Welles shale story is going to run in *Harper's*." About a half hour later, Ajemian called me and said the group had decided not to take any action against *Harper's*.

The following Wednesday, I learned that action would instead be taken against me. Hunt called me to his office and announced that "I'm sorry it has come to this, Chris." He understood why I felt I had to do what I did, but as managing editor he just could not let it pass. "You will have to vacate the premises within three weeks," he said.

And three weeks later, I was gone.

Index